Lecture Notes in Computer Science 13001

Ahmed Abdulkadir · Seyed Mostafa Kia ·
Mohamad Habes · Vinod Kumar ·
Jane Maryam Rondina · Chantal Tax ·
Thomas Wolfers (Eds.)

Machine Learning in Clinical Neuroimaging

4th International Workshop, MLCN 2021
Held in Conjunction with MICCAI 2021
Strasbourg, France, September 27, 2021
Proceedings

Springer

Editors
Ahmed Abdulkadir (ID)
University of Pennsylvania
Philadelphia, PA, USA

Mohamad Habes (ID)
The University of Texas Health Science
Center at San Antonio
San Antonio, TX, USA

Jane Maryam Rondina (ID)
University College London
London, UK

Thomas Wolfers (ID)
University of Oslo
Oslo, Norway

Seyed Mostafa Kia (ID)
Donders Institute
Nijmegen, The Netherlands

Vinod Kumar (ID)
Max Planck Institute for Biological
Cybernetics
Tübingen, Germany

Chantal Tax (ID)
University Medical Center Utrecht
Utrecht, The Netherlands

Cardiff University Brain Research Imaging
Centre (CUBRIC)
Cardiff, UK

ISSN 0302-9743   ISSN 1611-3349 (electronic)
Lecture Notes in Computer Science
ISBN 978-3-030-87585-5   ISBN 978-3-030-87586-2 (eBook)
https://doi.org/10.1007/978-3-030-87586-2

LNCS Sublibrary: SL6 – Image Processing, Computer Vision, Pattern Recognition, and Graphics

This Springer imprint is published by the registered company Springer Nature Switzerland AG
The registered company address is: Gewerbestrasse 11, 6330 Cham, Switzerland

Preface

Methodological developments in neuroimaging analysis contribute to the progress in clinical neurosciences. In specific domains of academic image analysis, impressive strides were made thanks to modern machine learning and data analysis methods such as deep artificial neural networks. The initial success in academic applications of complex neural networks started a wave of studies through the neuroimaging research field. Deep learning is now complementing more traditional machine learning as a tool for image and data analysis. It is our view that incorporating interdisciplinary domain knowledge into the machine learning models is critical to answer challenging clinically relevant research questions in the field of clinical neuroscience that eventually will translate to clinical routine. With this workshop, we aimed at creating an intellectual playing field for clinicians and machine learning experts alike to share and discuss knowledge at the interface between machine learning and clinical application.

The 4th International Workshop on Machine Learning in Clinical Neuroimaging (MLCN 2021) was held as a satellite event of the 24th International Conference on Medical Imaging Computing and Computer-Assisted Intervention (MICCAI 2021) to foster a scientific dialog between experts in machine learning and clinical neuroimaging. The call for papers was published on April 30, 2021, and the submission window closed on July 5, 2021. Each submitted manuscript was reviewed by three members of the Program Committee in a double-blindd review process. The accepted manuscripts contained in this proceedings presented a methodologically sound, novel, and thematically fitting contribution to the field of clinical neuroimaging, and were presented and discussed by the authors at the virtual MLCN workshop. The contributions studied *in vivo* structural and functional magnetic resonance imaging data. Several accepted submissions were concerned with computational anatomy involving a wide range of methods including supervised image segmentation, registration, classification, anomaly detection, and generative modeling. Network analysis and time series were other topical branches of the workshop contributions in which a wide variety of methods were employed and developed including dictionary learning, graph neural networks, and space-time convolutional neural networks. The fields of applications were as diverse as the methods. They included detection and modeling of abnormal cortical folding patterns and simulation of brain atrophy, mapping histology to ex vivo imaging, mapping functional cortical regions, and mapping structural with functional connectivity graphs. The methodological developments pushed the boundaries of clinical neuroscience image analysis with fast algorithms for complex and accurate descriptors of structure, function, or the combination of multiple modalities.

This workshop was made possible by a devoted community of authors, Program Committee, Steering Committee, and workshop participants. We thank all creators and attendees for their valuable contributions.

September 2021

Ahmed Abdulkadir
Mohamad Habes
Seyed Mostafa Kia
Vinod Kumar
Jane Maryam Rondina
Chantal Tax
Thomas Wolfers

Organization

Steering Committee

Christos Davatzikos	University of Pennsylvania, USA
Andre Marquand	Donders Institute, The Netherlands
Jonas Richiardi	Lausanne University Hospital, Switzerland
Emma Robinson	King's College London, UK

Organizing Committee

Ahmed Abdulkadir	University of Pennsylvania, USA
Mohamad Habes	University of Texas Health Science Center at San Antonio, USA
Seyed Mostafa Kia	University Medical Center Utrecht, The Netherlands
Vinod Kumar	Max Planck Institute for Biological Cybernetics, Germany
Jane Maryam Rondina	University College London, UK
Chantal Tax	University Medical Center Utrecht, The Netherlands
Thomas Wolfers	NORMENT, Norway

Program Committee

Mohammed Al-Masni	Yonsei University, South Korea
Andre Altman	University College London, UK
Pierre Berthet	University of Oslo, Norway
Özgün Çiçek	University of Freiburg, Germany
Richard Dinga	Donders Institute, The Netherlands
Charlotte Fraza	Donders Institute, The Netherlands
Pouya Ghaemmaghami	Concordia University, Canada
Francesco La Rosa	Ecole Polytechnique Fédérale de Lausanne, Switzerland
Sarah Lee	Amallis Consulting, UK
Hangfan Liu	University of Pennsylvania, USA
Emanuele Olivetti	Fondazione Bruno Kessler, Italy
Pradeep Reddy Raamana	University of Toronto, Canada
Saige Rutherford	University of Michigan, USA
Hugo Schnack	University Medical Center Utrecht, The Netherlands
Haochang Shou	University of Pennsylvania, USA
Haykel Snoussi	University of Texas Health Science Center at San Antonio, USA
Sourena Soheili Nezhad	Radboud University Medical Center, The Netherlands
Rashid Tanweer	University of Pennsylvania, USA
Petteri Teikari	University College London, UK

Erdem Varol   Columbia University, USA
Matthias Wilms   University of Calgary, Canada
Tianbo Xu   University College London, UK
Mariam Zabihi   Radboud University Medical Center, The Netherlands

Contents

Brain Networks and Time Series

Computational Anatomy

Unfolding the Medial Temporal Lobe Cortex to Characterize Neurodegeneration Due to Alzheimer's Disease Pathology Using Ex vivo Imaging

Sadhana Ravikumar[1]([✉]), Laura Wisse[2], Sydney Lim[1], David Irwin[1], Ranjit Ittyerah[1], Long Xie[1], Sandhitsu R. Das[1], Edward Lee[1], M. Dylan Tisdall[1], Karthik Prabhakaran[1], John Detre[1], Gabor Mizsei[1], John Q. Trojanowski[1], John Robinson[1], Theresa Schuck[1], Murray Grossman[1], Emilio Artacho-Pérula[3], Maria Mercedes Iñiguez de Onzoño Martin[3], María del Mar Arroyo Jiménez[3], Monica Muñoz[3], Francisco Javier Molina Romero[3], Maria del Pilar Marcos Rabal[3], Sandra Cebada Sánchez[3], José Carlos Delgado González[3], Carlos de la Rosa Prieto[3], Marta Córcoles Parada[3], David Wolk[1], Ricardo Insausti[3], and Paul Yushkevich[1]

[1] University of Pennsylvania, Philadelphia, USA
sadhanar@seas.upenn.edu
[2] Department of Diagnostic Radiology, Lund University, Lund, Sweden
[3] University of Castilla La Mancha, Albacete, Spain

Abstract. Neurofibrillary tangle (NFT) pathology in the medial temporal lobe (MTL) is closely linked to neurodegeneration, and is the early pathological change associated with Alzheimer's Disease (AD). In this work, we investigate the relationship between MTL morphometry features derived from high-resolution *ex vivo* imaging and histology-based measures of NFT pathology using a topological unfolding framework applied to a dataset of 18 human postmortem MTL specimens. The MTL has a complex 3D topography and exhibits a high degree of inter-subject variability in cortical folding patterns which poses a significant challenge for volumetric registration methods typically used during MRI template construction. By unfolding the MTL cortex, the proposed framework explicitly accounts for the sheet-like geometry of the MTL cortex and provides a two-dimensional reference coordinate space which can be used to implicitly register cortical folding patterns across specimens based on distance along the cortex despite large anatomical variability. Leveraging this framework in a subset of 15 specimens, we characterize the associations between NFTs and morphological features such as cortical thickness and surface curvature and identify regions in the MTL where patterns of atrophy are strongly correlated with NFT pathology.

Keywords: Medial temporal lobe · *Ex vivo* MRI · Cortical unfolding

---

Electronic supplementary material The online version of this chapter (https://doi.org/10.1007/978-3-030-87586-2_1) contains supplementary material, which is available to authorized users.

1 Introduction

The medial temporal lobe (MTL) is an essential component of the human memory system and the earliest region of the cortex affected by tau neurofibrillary tangles (NFT); a hallmark pathology associated with Alzheimer's Disease (AD). The accumulation of NFT pathology in the brain is closely linked to neurodegeneration and cognitive decline [1, 2]. According to studies by Braak and Braak [1], the spread of NFTs through the brain follows a characteristic pattern, with early manifestations observed in a specific region of the MTL surrounding the border between the lateral part of the entorhinal cortex (ERC) and the transentorhinal cortex (which corresponds to Brodmann area (BA) 35). The NFTs then spread further into the ERC before emerging in the hippocampus.

Measurements of neurodegeneration in MTL subregions can be derived using structural magnetic resonance imaging (MRI) and have been shown to be sensitive to changes during the early stages of AD [3]. However, the specificity of these measurements to AD is limited by the fact that aging, other neurodegenerative pathologies such as TDP-43, and vascular disease are frequently comorbid in patients with AD, and also cause structural changes in the MTL. Recent studies suggest that compared to NFT pathology, these concomitant pathologies are associated with different patterns of neurodegeneration within the MTL [4]. Therefore, improved characterization of the relationship between MTL neurodegeneration and NFT pathology could lead to the discovery of atrophy patterns that are strongly associated with NFT burden specifically, thereby contributing towards the development of in vivo imaging biomarkers for neurodegeneration that are more sensitive to longitudinal change in the presence of early AD.

Here, we study the relationship between MTL morphometry measures derived from high-resolution *ex vivo* imaging and histology-based measures of NFT pathology in a dataset of 15 human MTL specimens. The MTL has a complex topography and exhibits a high degree of anatomical variability, which poses a significant challenge during group-wise analyses that rely on creating a statistical reference space across all subjects. Typically, volumetric deformable registration is used to align individual anatomies and creates a reference space in the form of an average-shaped template of the anatomical region of interest [5]. However, when applied to the MTL, deformable registration can result in collapsing of different morphologies of the collateral sulcus, a basic landmark in the MTL [6, 7]. The variability in depth and shape of this sulcus can limit our ability to accurately infer the locations of MTL subregions such as BA35 and BA36 which are located along the collateral sulcus. A promising alternative is to use a surface-based approach which explicitly accounts for cortical folding patterns. In fact, *ex vivo* studies suggest that the locations of subregion borders depend on distance along the flattened cortical surface, indicating that MTL topology is an important consideration when studying inter-individual differences in MTL structure [8]. Therefore, in this work, we aim to investigate the relationship between NFT pathology and cortical thickness in a flattened space. Additionally, we seek to better understand the features of MTL morphometry driving the accumulation of NFT pathology by comparing our findings to the results of regional thickness analyses performed after alignment of cortical folding patterns using both surface-based and volumetric registration [7].

Existing tools such as FreeSurfer [9], developed for flattening the cortical surface *in vivo*, are not easily applicable to ultra-high-resolution ex vivo MRI. Instead, we customize the topological framework developed by DeKraker et al. [10] to create a two-dimensional (2D) unfolded representation of the extra-hippocampal MTL cortex, which includes the ERC, BA35, BA36 and the parahippocampal cortex (PHC). This unfolded coordinate space can be used to index locations of the extrahippocampal MTL cortex in 2D based on their distance from the hippocampus and thus provides implicit registration between specimens despite differences in sulcal depth and folding patterns. Leveraging this framework, we identify regions of the MTL where atrophy correlates most strongly with NFT burden. In an exploratory analysis, we show that groupwise registration of MTL sulcal patterns across specimens can be performed using surface curvature for local shape comparison.

2 Ex-vivo Imaging Dataset

2.1 Specimen Preparation and Imaging

Intact *ex vivo* brain bank specimens of the MTL were obtained from 18 donors (12 males, aged 45–93) from the University of Pennsylvania (UPenn) and the University of Castilla-La Mancha (UCLM) in Spain. Human brain specimens were obtained in accordance with the UPenn Institutional Review Board guidelines, and the Ethical Committee of UCLM. Where possible, pre-consent during life and, in all cases, next-of-kin consent at death was given. Following 4+ weeks of fixation, the MTL blocks were imaged on a Varian 9.4 T animal scanner at a $0.2 \times 0.2 \times 0.2$ mm^3 resolution using a T2-weighted, multislice spin echo sequence (TE $= 9330$ ms, TR $= 23$ ms). Due to gradient distortions in the 9.4 T scanner, as part of the post-processing, all of the scans had to be warped to correct for differences between the scanner coordinate frame and physical coordinate frame. Linear scaling factors for this transformation were derived using a 3D printed phantom [11]. Following MRI scanning, the specimens underwent serial histological processing. Specimens were cut into 2 cm blocks using custom molds that were 3D printed to fit each MTL specimen, frozen and sectioned at 50 μm intervals. Every 10th section was stained for cytoarchitecture (Nissl stain) and in 15 specimens, every 20th section was prepared for immunohistochemistry (IHC) with the anti-tau AT8 antibody and Nissl counterstain. Sections were mounted on 7.5 cm 5 cm slides and digitally scanned at 20X resolution. For each block, the scanned sections were reconstructed in 3D and aligned to MRI space using a custom deformable 3D registration pipeline.

2.2 Quantitative NFT Burden Maps from Histology

For the 15 specimens with anti-tau IHC, "heat maps" quantifying the burden of NFT pathology on each of the anti-tau IHC sections were generated using a weakly supervised deep learning algorithm as described in [12]. Given an input patch extracted from a histology slide, the network outputs a spatial heatmap indicating the intensity of tangles at each location. The automated NFT burden measures generated by the network were shown to be consistent with manual NFT counts and semi-quantitative ratings of NFT severity provided by an expert neuropathologist. For each specimen, the trained network was applied to all anti-tau whole-slide IHC images and the resulting heatmaps were reconstructed into a 3D volume and transformed into the space of the 9.4T MRI. Further details of the histology protocol, the approach for 3D reconstruction and matching of histology to MRI and NFT mapping are provided in [12].

2.3 Histology-Guided MTL Subregion Segmentations

In 11 specimens, the MTL subregions were labeled in *ex vivo* MRI space based on cytoarchitectural features derived from the serial histology images (Nissl stain). These histology guided segmentations are performed manually on each MRI slice and are highly labor intensive. Therefore, they are currently only available in a subset of cases. First, the boundaries between hippocampal subfields and extrahippocampal subregions (ERC, BA35, BA36, area TE and the PHC (areas TF and TH)) were identified in the histology images. Following histology reconstruction and registration to MRI space, the boundary annotations were mapped into 3D MRI space and overlaid on the co-registered MRI and histology images. Guided by the boundary annotations, the subfield segmentations were manually traced in 3D MRI space (Fig. 1C). Note that for each specimen, small gaps in the segmentation may exist between histology blocks.

3 Methods

3.1 Overview of Topological Unfolding Framework

To unfold the extra-hippocampal MTL, we applied the framework developed by DeKraker et al. [10, 13] which imposes a curvilinear coordinate system on the cortex by solving a set of Laplace's equations along segmentations of the gray matter. DeKraker et al. propose unfolding the hippocampus by computing potential field gradients along the anterior-posterior and proximal-distal directions of the cortex. This is done by defining additional boundary conditions at the anterior, posterior, proximal and distal ends of the region of interest and solving Laplace's equation for three sets of boundary conditions, $\nabla^2\varphi_{AP} = 0$, $\nabla^2\varphi_{PD} = 0$, $\nabla^2\varphi_{IO} = 0$ in the anterior-posterior (AP), proximal-distal (PD) and laminar (IO) directions respectively. We focus on unfolding the extra-hippocampal MTL by first segmenting the extra-hippocampal region over which the potential field is defined, labelling the boundary surfaces, and then solving the set of three Laplace's equations to generate a potential field in each direction.

3.2 Segmentation of the Outer MTL Boundary in Ex vivo MRI

In the *ex vivo* MRI scans of each of the 18 specimens, separate labels were used to segment the MTL gray matter and six boundary surfaces. The MTL gray matter was segmented using a semi-automated interpolation approach which combines inter-slice interpolation [14] and manual editing to reduce the manual effort needed to generate the segmentations. In the anterior region of MTL cortex, the gray matter segmentation extends until the medial bank of the occipitotemporal sulcus. In the posterior region of

MTL cortex, heuristically defined as starting 6 mm after the end of the hippocampus head, the gray matter segmentation only extends until the fundus of the collateral sulcus. This difference in lateral boundaries between the two regions introduces a discontinuity in the extent of unfolded tissue. Figure 1A shows an example MTL segmentation with boundary labels.

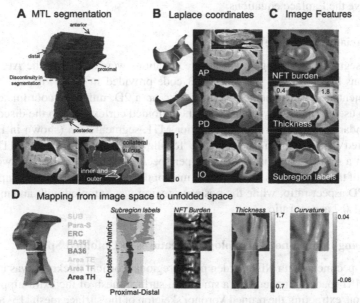

Fig. 1. Illustration of the topological unfolding framework applied to the extra-hippocampal MTL in an example specimen. (A) 3D reconstruction and coronal view of the semi-automated segmentation of the extra-hippocampus (red) and the six boundary labels used for solving Laplace's equation along the anterior-posterior (AP), proximal-distal (PD) and inner-outer (IO) directions. The inner and outer boundary labels are marked in the cross-sectional view. (B) Coronal view and mid-surface model of the Laplacian solutions in each direction. (C) Coronal view of the NFT burden map, cortical thickness measurements (in mm) and MTL subregion labels in native MRI space. (D) Subregion labels shown in native and unfolded space along with unfolded representations of NFT burden, cortical thickness, and mean curvature along the cortical surface. (Color figure online)

Prior to segmentation, the MRI scan for each specimen was re-oriented such that the long-axis of the hippocampus aligned with the anterior-poster direction. Therefore, the AP boundary labels were obtained by dilating the MTL gray matter segmentation in the anterior-posterior dimension. Since the unfolding framework focuses on the extra-hippocampal MTL, we defined the hippocampus as the proximal boundary. We note that part of the subiculum and parasubiculum is included in the unfolded tissue (they form the medial boundary, particularly for the PHC). The distal boundary was obtained by manually labelling the voxels bordering the lateral extent of the MTL gray matter segmentation on each slice. Lastly, the inner white matter and outer pial surfaces were labelled using a semi-automated approach which involved dilating the MTL segmentation in the coronal plane and re-labelling the voxels falling within the background (identified by thresholding the MRI scan) with a different label to separate the inner and outer boundaries. Following this process, each of the completed segmentations were visually evaluated and any errors were manually corrected. Figure 1A shows a coronal view and 3D reconstruction of an example MTL segmentation including the boundaries used to solve the Laplace equations.

3.3 Laplacian Coordinate System

Given the segmentation image, the Laplace equations were solved in the AP, PD, and IO directions by modifying the MATLAB code provided in [13] (Fig. 1B). The AP and PD potential field gradients together make up a 2D, unfolded coordinate system that can be used to index any point along the unfolded cortex. Due to the discontinuity at the boundary of the anterior and posterior MTL segmentation (shown in Fig. 1A), we computed Laplace solutions separately for the anterior and posterior MTL cortex resulting in a set of two coordinate maps per specimen. To reflect the real-world size and extent of the two regions, the unfolded map of the anterior region was sampled with a 1:1 AP:PD aspect ratio, while the posterior region was sampled with an empirically estimated, 1:0.7 aspect ratio.

3.4 Mapping Image and Morphological Features to Unfolded Space

For the 15 specimens with NFT burden maps, regional cortical thickness was measured in native MRI space by generating a smoothed surface mesh of the extra-hippocampal segmentation, extracting the pruned Voronoi skeleton of the surface mesh [15] and computing twice the distance between each vertex and the closest point on the skeleton. The NFT heatmaps, subregion labels, and thickness measurements (shown in Fig. 1C), were transformed from native MRI space to unfolded space as follows: the mid-surface of the extra-hippocampal MTL was extracted by interpolating the 3D native-space coordinates corresponding to the unfolded points at a laminar potential of 0.5 from the inner and outer surfaces. Delaunay triangulation was used to perform scattered interpolation [16]. Image features were then sampled from the nearest-neighbor interpolated location in MRI space for each point along the mid-surface. Additionally, mean curvature was estimated at each vertex along the 0.5-level mid-surface of the cortex using the patchcurvature() function in MATLAB. Gaussian smoothing was then applied to the thickness and curvature maps for each specimen in a reparametrized unfolded space that reflects the real-world distances between points (Supplementary Table 1) [13]. Figure 1D shows examples of the four image features mapped to unfolded space. Additionally, the unfolded feature maps for all 18 specimens are shown in Supplementary Fig. 1.

4 Experiments and Results

4.1 Consensus MTL Subregion Segmentation in Unfolded Coordinate Space

An average MTL subregion segmentation was generated in the unfolded space by performing voxel-wise majority voting among the subregion segmentations of the subset

of 11 specimens with histology-based annotations. When obtaining the consensus segmentation, we incorporated slight regularization using a Markov Random Field prior to
smooth the boundaries between labels and provide continuity at voxels where little data is
available (due to gaps in the histology segmentations) (Fig. 2A and Supplementary Table
1). For each pair of specimens, we computed the generalized Dice coefficient (GDSC)
between their multi-label segmentations in unfolded space, only including points that are
labeled in both specimens [17]. The average GDSC across the 11 specimens is 0.62 ±
0.09. This is higher than the average GDSC of 0.57 ± 0.07 obtained when registering
specimens using a volumetric deformable registration pipeline [7], suggesting improved
alignment of the MTL cortex of individual specimens in unfolded space.

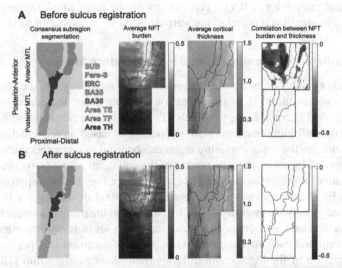

Fig. 2. Comparison of the MTL segmentation, results of the regional thickness analysis, and
average maps of cortical thickness and NFT burden, visualized in unfolded space, before and
after sulcus registration using mean curvature. The consensus MTL subregion segmentation is
derived from serial histology in 11 specimens. The statistical maps show regions where significant
correlations were observed between cortical thickness and the 90th percentile of NFT burden in
BA35, using the Spearman rank correlation model with age as a covariate, in a dataset of 15
specimens. Point-wise correlations were considered significant after false discovery rate correction
($p < 0.1$). The thickness and NFT burden maps shown in the third and fourth column respectively,
represent the average values computed across 15 specimens. In each case, the boundaries of the
consensus segmentation are overlaid in black.

4.2 Correlating NFT Burden with MTL Neurodegeneration

To characterize the effects of NFT pathology on MTL thickness, for each of the 15 specimens with NFT burden heatmaps we first computed a summary measure of NFT severity,
defined as the 90th percentile of NFT burden across all points in the unfolded map that
fall within BA35. We used BA35 to compute the summary measure since it contains

the transentorhinal cortex, the first cortical site affected by NFTs in AD [1]. Since only five of the specimens in the dataset have both NFT burden maps and histology-based subregion segmentations, the consensus segmentation was used to determine the boundary of BA35 in all specimens. We then performed a statistical analysis to investigate the relationship between the NFT summary measure and thickness (standardized across subjects) at each location in the unfolded space using the partial Spearman rank correlation model with age as a covariate. Point-wise correlations were considered significant after false discovery rate (FDR) correction ($p < 0.1$). P-values and the corresponding FDR threshold are plotted in Supplementary Fig. 3. Due to missing data along the borders of the thickness maps, 10% of values along the anterior, posterior, proximal and distal edge were not included in the analysis. As seen in Fig. 2A, strong correlations were observed in the ERC and the border of BA35, consistent with the early Braak regions, and parts of BA36 [1]. No significant correlations were detected in the posterior MTL.

4.3   Surface-Based Registration Using Mean Curvature Maps

In an exploratory analysis, we were interested in disentangling whether the distribution of NFTs within the MTL is dependent on distance along the cortex or merely a consequence of MTL morphology, with NFTs tending to accumulate within cortical folds. While the unfolding framework aligns specimens based on relative distance from a boundary surface, the mean curvature maps can be used to align specimens based on sulcal patterns and location since the fundus of the collateral sulcus is visible as regions of high curvature in the unfolded maps (Supplementary Fig. 1). This approach is analogous to the FreeSurfer surface-based registration method developed for in vivo MRI [9]. We applied groupwise intensity-based registration to the mean curvature maps of all 18 specimens to create an average curvature map and a set of transformations between the average map and each specimen [5]. Groupwise registration was performed using an implementation of the log domain diffeomorphic demons algorithm [18] included in the "Greedy" package and involved iteratively alternating between highly regularized, deformable registration of the individual curvature maps to an average template using the sum of squared differences metric and updating the template by averaging the registered maps (Supplementary Table 1). Supplementary Fig. 2 shows the average curvature map before and after registration. We observe that the regions of high curvature are better defined following registration, suggesting improved alignment of the collateral sulcus between specimens. To test the relationship between NFT burden and cortical thickness in this space, we repeated the pointwise thickness analysis after first mapping the subregion labels, NFT burden maps and thickness maps to the normalized space and re-computing the consensus subregion segmentation and NFT summary measures. As shown in Fig. 2B (and Supplementary Fig. 3), no significant correlations are observed following sulcus-based registration, suggesting that patterns of tau distribution and neurodegeneration are perhaps more dependent on distance along the cortex than sulcal folding patterns. This finding is consistent with the result obtained when we performed pointwise regional thickness analysis in the space of a 3D MTL template generated using volumetric deformable registration (Supplementary Fig. 3) [7]. While the expected effect of MTL morphology on NFT pathology is unclear, this result is also consistent with the findings reported in a recent study by Arena et al. that assessed the preferential distribution of tau pathology towards sulcal depths in the context of chronic traumatic encephalopathy (CTE) and AD and found that NFTs showed a more uniform distribution along the cortex when compared to astroglial tau pathology [19].

5 Conclusions

We present a topological unfolding framework applied to the extrahippocampal MTL cortex, using ex vivo MR imaging of a sizable collection of human MTL specimens (n = 18). This approach allows us to visualize, for the first time, the distribution of extrahippocampal subregions and NFT pathology in an unfolded space and analyze the effects of NFT burden on MTL neurodegeneration while explicitly accounting for the complex topology of the MTL. Our result suggesting that the association between NFT burden and cortical thickness is weakened following alignment of sulcal patterns motivates further work in a flattened space with a larger dataset. While we show that the unfolding framework provides a valuable tool for detailed investigation of MTL neurodegeneration due to NFT pathology, in ongoing work, IHC for other common molecular pathologies is being performed in many of our specimens and the unfolding framework can be easily extended to investigate a different set of image features. Future work will focus on expanding the size of our dataset and extending the unfolding framework to the hippocampal subfields, which include early targets for NFT pathology. This will allow us to further refine our understanding of early AD and support the development of better AD biomarkers.

References

1. Braak, H., Braak, E.: Neuropathological staging of Alzheimer-related changes. Acta Neuropathol. **82**, 239–259 (1991). https://doi.org/10.1007/bf00308809
2. Hyman, B.T., et al.: National institute on aging–Alzheimer's association guidelines for the neuropathologic assessment of Alzheimer's disease. Alzheimer's Dement. **8**, 1–13 (2012). https://doi.org/10.1016/j.jalz.2011.10.007
3. Olsen, R.K., Palombo, D.J., Rabin, J.S., Levine, B., Ryan, J.D., Rosenbaum, R.S.: Volumetric analysis of medial temporal lobe subregions in developmental amnesia using high-resolution magnetic resonance imaging. Hippocampus **23**, 855–860 (2013). https://doi.org/10.1002/hipo.22153
4. Small, S.A., Schobel, S.A., Buxton, R.B., Witter, M.P., Barnes, C.A.: A pathophysiological framework of hippocampal dysfunction in ageing and disease (2011). https://doi.org/10.1038/nrn3085
5. Joshi, S., Davis, B., Jomier, M., Gerig, G.: Unbiased diffeomorphic atlas construction for computational anatomy. NeuroImage Neuroimage (2004). https://doi.org/10.1016/j.neuroimage.2004.07.068
6. Xie, L., et al.: Automatic clustering and thickness measurement of anatomical variants of the human perirhinal cortex. In: Golland, P., Hata, N., Barillot, C., Hornegger, J., Howe, R. (eds.) MICCAI 2014. LNCS, vol. 8675, pp. 81–88. Springer, Cham (2014). https://doi.org/10.1007/978-3-319-10443-0_11
7. Ravikumar, S., et al.: Building an ex vivo atlas of the earliest brain regions affected by Alzheimer's disease pathology. In: Proceedings - International Symposium on Biomedical Imaging (2020). https://doi.org/10.1109/ISBI45749.2020.9098427

8. Ding, S.L., Van Hoesen, G.W.: Borders, extent, and topography of human perirhinal cortex as revealed using multiple modern neuroanatomical and pathological markers. Hum. Brain Mapp. **31**, 1359–1379 (2010). https://doi.org/10.1002/hbm.20940

9. Fischl, B., Sereno, M.I., Tootell, R.B.H., Dale, A.M.: High-resolution intersubject averaging and a coordinate system for the cortical surface. Hum. Brain Mapp. **8**, 272–284 (1999). https://doi.org/10.1002/(SICI)1097-0193(1999)8:4%3c272::AID-HBM10%3e3.0.CO;2-4

10. DeKraker, J., Ferko, K.M., Lau, J.C., Köhler, S., Khan, A.R.: Unfolding the hippocampus: an intrinsic coordinate system for subfield segmentations and quantitative mapping. Neuroimage **167**, 408–418 (2018). https://doi.org/10.1016/j.neuroimage.2017.11.054

11. Adler, D.H., et al.: Characterizing the human hippocampus in aging and Alzheimer's disease using a computational atlas derived from ex vivo MRI and histology. Proc. Natl. Acad. Sci. U.S.A. **115**, 4252–4257 (2018). https://doi.org/10.1073/pnas.1801093115

12. Yushkevich, P.A., et al.: Three-dimensional mapping of neurofibrillary tangle burden in the human medial temporal lobe. Brain **139**, 16–17 (2021). https://doi.org/10.1093/BRAIN/AWAB262

13. DeKraker, J., Lau, J.C., Ferko, K.M., Khan, A.R., Köhler, S.: Hippocampal subfields revealed through unfolding and unsupervised clustering of laminar and morphological features in 3D BigBrain. Neuroimage **206** (2020). https://doi.org/10.1016/j.neuroimage.2019.116328

14. Ravikumar, S., Wisse, L., Gao, Y., Gerig, G., Yushkevich, P.: Facilitating manual segmentation of 3D datasets using contour and intensity guided interpolation. In: 2019 IEEE 16th International Symposium on Biomedical Imaging (ISBI 2019), pp. 714–718 (2019)

15. Ogniewicz, R.L., Kübler, O.: Hierarchic Voronoi skeletons. Pattern Recogn. **28**, 343–359 (1995). https://doi.org/10.1016/0031-3203(94)00105-U

16. Amidror, I.: Scattered data interpolation methods for electronic imaging systems: a survey. J. Electron. Imaging **11**, 157 (2002). https://doi.org/10.1117/1.1455013

17. Crum, W.R., Camara, O., Hill, D.L.G.: Generalized overlap measures for evaluation and validation in medical image analysis. IEEE Trans. Med. Imaging. **25**, 1451–1461 (2006). https://doi.org/10.1109/TMI.2006.880587

18. Vercauteren, T., Pennec, X., Perchant, A., Ayache, N.: Symmetric log-domain diffeomorphic registration: a demons-based approach. In: Metaxas, D., Axel, L., Fichtinger, G., Székely, G. (eds.) MICCAI 2008. LNCS, vol. 5241, pp. 754–761. Springer, Heidelberg (2008). https://doi.org/10.1007/978-3-540-85988-8_90

19. Arena, J.D., et al.: Astroglial tau pathology alone preferentially concentrates at sulcal depths in chronic traumatic encephalopathy neuropathologic change. Brain Commun. **2** (2020). https://doi.org/10.1093/BRAINCOMMS/FCAA210

Distinguishing Healthy Ageing from Dementia: A Biomechanical Simulation of Brain Atrophy Using Deep Networks

Mariana Da Silva[1](✉), Carole H. Sudre[1,2,3], Kara Garcia[4], Cher Bass[1,5], M. Jorge Cardoso[1], and Emma C. Robinson[1]

[1] School of Biomedical Engineering and Imaging Sciences, King's College London, London, UK
mariana.da_silva@kcl.ac.uk
[2] MRC Unit for Lifelong Health and Ageing at UCL, University College London, London, UK
[3] Centre for Medical Image Computing, Department of Computer Science, University College London, London, UK
[4] Department of Radiology and Imaging Sciences, School of Medicine, Indiana University, Bloomington, USA
[5] Panakeia Technologies, London, UK

Abstract. Biomechanical modeling of tissue deformation can be used to simulate different scenarios of longitudinal brain evolution. In this work, we present a deep learning framework for hyper-elastic strain modelling of brain atrophy, during healthy ageing and in Alzheimer's Disease. The framework directly models the effects of age, disease status, and scan interval to regress regional patterns of atrophy, from which a strain-based model estimates deformations. This model is trained and validated using 3D structural magnetic resonance imaging data from the ADNI cohort. Results show that the framework can estimate realistic deformations, following the known course of Alzheimer's disease, that clearly differentiate between healthy and demented patterns of ageing. This suggests the framework has potential to be incorporated into explainable models of disease, for the exploration of interventions and counterfactual examples.

Keywords: Deep learning · Biomechanical modelling · Neurodegeneration · Disease progression

1 Introduction

Alzheimer's Disease (AD) is neurodegenerative condition characterized by progressive and irreversible death of neurons, which manifests macroscopically on structural magnetic resonance images (MRI) as progressive tissue loss or atrophy. While, cross-sectionally the progression of the disease is well documented - presenting with disproportionate atrophy of the hippocampus, medial temporal,

A. Abdulkadir et al. (Eds.): MLCN 2021, LNCS 13001, pp. 13–22, 2021.
https://doi.org/10.1007/978-3-030-87586-2_2

and posterior temporoparietal cortices [7,17], relative to age matched controls - in reality disease progression is heterogeneous across individuals and may be categorised into subtypes [8]. Historically, this has meant that early stage AD has been challenging to diagnose from structural MRI changes alone [1,9,11,15].

Biomechanical models present an alternate avenue, in which rather than performing post-hoc diagnosis of AD from longitudinally acquired data, it instead becomes possible to build a forward model of disease, simulating different possible scenarios for progression [12,13]. Such models have been used broadly throughout the literature to simulate both atrophy and growth [18,20,22] and are usually based on hyperelastic strain models, implemented using finite element methods (FEM) [21] or finite difference methods (FDM) [12].

Accordingly, in this paper we propose a novel deep network for biomechanical simulation of brain atrophy, and seek to model differential patterns of atrophy following healthy ageing or AD. In this way our model parallels a growing body of deep generative, interpretable or explainable models of disease. This includes [3–5,14] which train generative models to deform [14] and/or change the appearance [3–5,14] of images, in such a way that it changes their class. By contrast, deep structural causal models such as [16], go further to support counterfactual models of disease progression, by associating demographic and phenotypic variables to imaging data, through variational inference on a causal graph.

One challenge with structural causal models is that they require prior hypothesis of a causal graph, defining the directions of influence of different parameters in the model. In this paper, we therefore take a more explicit approach to explainable modelling, training a hyper-elastic strain simulation of brain growth and atrophy, while building an explicit simulation of atrophy for different populations and time windows. This supports subject-specific interventions, simulating projections of brain atrophy, following differing diagnoses.

2   Methods

2.1   Data

Data used in this study were obtained from the Alzheimer's Disease Neuroimaging Initiative (ADNI) database[1]. A total of 1054 longitudinal MRI scans, collated from the ADNI1, ADNI2, ADNI-GO and ADNI3 studies, were used. All examples have at least 2 different T1-weighted scans, separated by at least 1 year (range 1–14 years). Accelerated MRI data was used for the subjects that don't have non-accelerated images for both time-points. The dataset includes 210 subjects diagnosed with AD, 677 subjects with Mild Cognitive Impairment (MCI), 67 subjects with Significant Memory Loss (SMC) and 92 cognitively normal (CN). From this, subjects were separated into 845 training datasets, 104 validation datasets, and 105 test datasets. An equal distribution of the 4 disease classes was ensured in each set.

---

[1] http://adni.loni.usc.edu/.

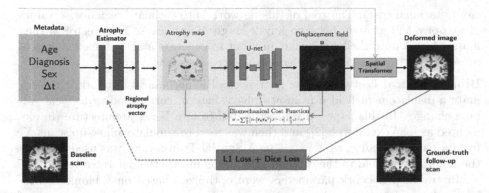

Fig. 1. Model architecture: the biomechanical model estimates a deformation field from a prescribed atrophy map corresponding to local volume changes; the atrophy estimator predicts region-wise atrophy values based on demographics and time. In this work, we train the two networks in different stages: the model highlighted in yellow is pre-trained based on a biomechanical cost function (green arrows); the atrophy estimator is trained based on the similarity metrics between simulated and true follow-up image (blue arrows). At inference time, the model estimates a follow-up scan based on metadata and a baseline image alone (the testing procedure can be identified by the black and grey arrows). (Color figure online)

2.2 Preprocessing

MRI images were segmented into cerebrospinal fluid (CSF), white matter (WM), gray matter (GM), deep gray matter (DGM) and cerebellum using NiftySeg[2]. The images were parcellated into 138 regions (NeuroMorph parcellations) using the geodesic information flow (GIF) algorithm [6]. We then generate a less granular parcellation of 27 regions that includes the separate cortical lobes, ventricular system and hippocampus, which we use in the model and for our analysis. T1 images were skull stripped based on the segmentations, then resampled to MNI space with rigid registration using FSL's FLIRT [10]. Data were normalised into the range 0–1 using histogram normalization, based on data from a target subset of 50 subjects.

2.3 Model Overview

Figure 1 offers an overview of the model and the training procedure. The model consists of two networks: an Atrophy Estimator and a Biomechanical Network.

The Atrophy Estimator is a two-hidden layer (32 and 64 nodes) multi-layer perceptron (MLP) which takes as input 4 demographic variables: biological age (at the time of the first scan - normalized), sex, disease class (CN, SMC, MCI or AD) and time-interval between scans (Δt). It predicts as output a tensor of size 27, which corresponds to a predicted atrophy or growth value for each region of the brain (using the less granular parcellations, in order to reduce

---

[2] http://github.com/KCL-BMEIS/NiftySeg.

computational cost). The goal of this network is to estimate region-wise values of atrophy and growth, between any 2 longitudinal scans. This vector is then mapped back onto the label image, in order to generate a 3D volumetric map of prescribed atrophies, piece-wise constant, across regions.

Biomechanical Network: The goal of the Biomechanical Network is to estimate a displacement field $\mathbf{u}$ from atrophy values, a, corresponding to local volume changes. In this paper, $\mathbf{u}$ was estimated from a U-net architecture (implemented as for VoxelMorph [2]) and then was used to simulate follow-up scans X, from a baseline scan x, as $X = x + \mathbf{u}$. A Spatial Transformer was used to apply the deformation field to the original grid and compute the deformed image.

In training, network parameters were optimized based on a biomechanics-inspired cost function. Following the convention used in modelling growth of biological tissues [19,23], we model the brain as a Neo-Hookean material and minimise the strain energy density, W:

$$W = \sum \frac{\mu}{2} \left[Tr\left(\mathbf{F_K}\mathbf{F_K}^\mathrm{T} \right) J^{-2/3} - 3 \right] + \frac{K}{2}(J - 1)^2 \tag{1}$$

Here, $J = \det(\mathbf{F_K})$, and the elastic deformation $\mathbf{F_K}$ is responsible for driving equilibrium. This is given by $\mathbf{F_K} = \mathbf{F} \cdot \mathbf{G}^{-1}$, where $\mathbf{F} = \nabla\mathbf{u} + \mathbf{I}$ is the total deformation gradient and $\mathbf{G}$ is the applied growth, $\mathbf{G} = (a^{-1/3})\mathbf{I}$. a represents relative changes in volume, and we assume isotropic growth/atrophy. μ is the shear modulus and $K = 100\mu$ is the bulk modulus. We define $\mu = 1$ for pixels belonging to GM and WM, and set $\mu = 0.01$ for the CSF, which we model as a quasi-free tissue. As only the tissues inside the skull suffer deformation, we add a loss term to encourage zero displacement in the voxels outside of the CSF. We also minimize the displacement at the voxel corresponding to the centre of mass of the brain. The total cost function is:

$$\mathcal{L}_{Biomechanical} = W + \lambda_1 \sum \|\mathbf{u_{background}}\|^2 + \lambda_2 \|\mathbf{u_{center}}\|^2, \tag{2}$$

where λ_1, λ_2 are hyperparameters weighting the contribution of these terms.

2.4 Training and Evaluation

We train the two networks of our model in two separate stages:

Pre-training the Biomechanical Model: Here, ground-truth per-region atrophy maps were first calculated from the volume ratio between the 2 time-points for each of the original 138 NeuroMorph parcellations ($a_{ground-truth} = V_1/V_2$). These were then used to simulate a range of possible atrophy maps by sampling, for each region, from a uniform distribution of plausible atrophies (with range constrained between the min and max values of each population). In this way, the diversity of training samples seen by the model was increased.

At each iteration, the model was trained with either a subject-specific ground-truth atrophy or an atrophy pattern randomly sampled from these distributions. We note that the aim here is to train the network to estimate displacement fields from any reasonable value of prescribed growth or atrophy, rather than learn deformation patterns from the population. The biomechanical model was trained for 200 epochs using a mini-batch size of 6 and ADAM optimizer with a learning rate of 1×10^{-4}. Based on our previous experiments, we set $\lambda_1 = 10^{-1}$ and $\lambda_2 = 10^2$.

Atrophy Simulation: Subsequently, the atrophy estimator was trained to predict the atrophies from the subject demographics and time-window. To train this network, the MLP outputs are applied to the pre-trained biomechanical model to compute the corresponding displacement field, simulated image and simulated parcellations. We then update the weights of the MLP based on the average Soft Dice Loss across the 27 parcellations and the $\mathcal{L}_1$ loss between simulated follow-up scan and ground-truth follow-up scan. The total loss of this network is therefore given by:

$$\mathcal{L}_{MLP} = SoftDice + 0.1\mathcal{L}_1 \tag{3}$$

The network was trained for 50 epochs, with batch size = 3 and learning rate = 1×10^{-4}.

3   Experimental Methods and Results

3.1   Evaluation of Biomechanical Model

We apply the trained biomechanical model to the region-wise ground-truth volume change values of the 105 subjects of the test set. Figure 2 shows a representative example of a prescribed atrophy map, computed atrophy ($\det(\mathbf{F})$), simulated follow-up scan and corresponding ground-truth for a subject diagnosed with MCI, with a time-span between scans of 7 years.

We evaluate the performance of the network by comparing the prescribed atrophy maps with the computed atrophy using the Mean Squared Error (MSE), and compare the simulated images and segmentations to the ground-truth using MSE and dice overlap scores. In addition, we calculate the Absolute Symmetric Percentage Volume Change (ASPVC) between the simulated and ground-truth follow up images as in [13]. The objective is to show that the biomechanical network can estimate realistic deformation fields when provided with a specific atrophy map corresponding to local volume changes.

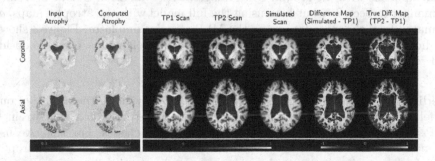

Fig. 2. Results of biomechanical model applied to a ground-truth atrophy map. The simulated follow-up scan shows atrophy of the ventricles and cortex that approximates the true difference map between the scans.

Table 1. Evaluation metrics (Mean and Standard Deviation) calculated over the 105 subjects of the test set, for the biomechanical model applied to the ground-truth atrophy maps.

	$\mathrm{MSE}_{atrophy}$	MSE_{Image}	Dice_{vent}	Dice_{cortex}	ASPVC_{vent}	ASPVC_{cortex}
Mean	1.27×10^{-4}	2.38×10^{-3}	0.901	0.760	2.6 %	4.1 %
Standard deviation	2.87×10^{-4}	1.65×10^{-3}	0.047	0.065	4.5 %	5.0 %

Table 1 shows the evaluation metrics over the test set, with focused analysis of the dice and ASPVC metrics for ventricles and cortex. We report high dice overlap and low ASPVC for the ventricular region across all subjects, showing that the model can accurately simulate the deformation patterns in the ventricles. Calculated ASPVC values are inside the range (2%–5%) of values reported in [13], which simulated atrophy using a FDM model. Lower values of dice overlap for the cortex region can be explained partially by registration differences between the two scans, which influence not only the comparison between simulated and ground-truth image, but also the "ground-truth" volume changes used as input to the model, that are calculated from the parcellations. Note throughout, that 0% volume change could only be expected for the images if the model was prescribed precise voxel-wise atrophy values.

3.2 Evaluation of Atrophy Estimation

Our next aim is to use this model to simulate patterns of atrophy according to different conditions, including the elapsed time between scans and the disease status. In this section, we show that: 1) our atrophy estimation model is capable of simulating follow-up scans consistent with the ground-truth; 2) the model can differentiate between atrophy for healthy, MCI and AD subjects; 3) the model can project forward in time.

Comparison with Ground-Truth: We start by evaluating our atrophy estimation model on the 105 subjects of the test set and compare the simulated follow-up scans with the ground-truth images. This evaluation is done in a similar manner to Sect. 3.1, but here using the full model with the atrophy maps, a, estimated from the metadata using the MLP atrophy simulator. Figure 3 shows the dice overlap between simulated and ground-truth follow-ups, over all considered regions.

Predicting Trajectories of Disease: In order to evaluate the ability of our model to differentiate between healthy aging, MCI and AD, we use our trained MLP to estimate atrophy patterns for the different classes. For this, we use the metadata from the 105 subjects of the test set and re-estimate the atrophy maps by intervening on input channel corresponding to the diagnosis class. We

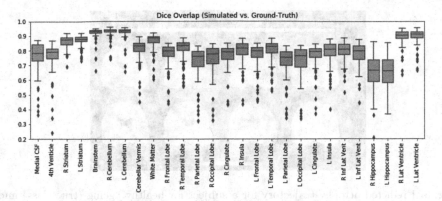

Fig. 3. Dice overlap scores between the simulated and ground-truth parcellations for each of the 27 regions, computed across the 105 subjects of the test set.

therefore calculate 4 atrophy maps for each subject (disease class = {CN, SMC, MCI, AD}), and keep the remaining metadata (age, sex, Δt) as the true values. Figure 4 shows the predicted atrophy values for the ventricles and hippocampus when intervening on the disease class.

We performed one-sided paired t-test analysis on the computed atrophies, and conclude that, for the ventricles, the model is able to predict statistically significant differences in atrophy distributions between all 4 diagnosis classes ($P < .001$). For the hippocampus, the model estimates atrophies that are significantly different when comparing CN vs MCI, CN vs. AD and MCI vs. AD ($P < .001$); the distributions for CN and SMC are not significantly different ($P = 0.89$).

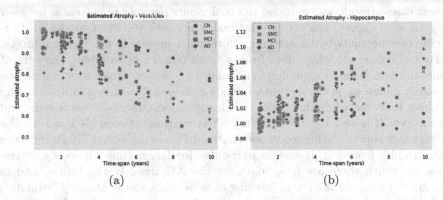

Fig. 4. Predicted atrophy values on (a) ventricles and (b) hippocampus when intervening on disease status. Values of $a < 1$ correspond to regional expansion and $a > 1$ correspond to shrinking.

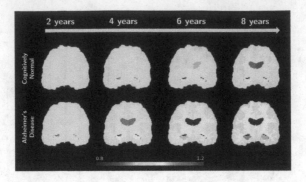

Fig. 5. Predicted atrophy trajectory for a subject for healthy ageing (true class) and in the presence of AD. Age at baseline = 82 years.

Finally, to show that the model can predict forward in time, we estimate the atrophy for the subjects of the test set for multiple time-spans (Δt = 2, 4, 6, and 8 years). Figure 5 shows the computed atrophy progression when considering healthy aging, and when changing the input class to Alzheimer's Disease. Comparing the trajectories for both cases, it is visible that the model predicts, as expected, larger values of atrophy across the brain tissue, including the ventricles and hippocampus.

4   Discussion and Future Work

The results presented here show that the proposed framework can be used to simulate structural changes in brain shape resulting from neurodegenerative disease, and differentiate between healthy and diseased atrophy patterns. In the present case, atrophy was encoded to predict only demographic trends; however our goal is to expand the atrophy estimator network to better model subject-specific heterogeneity by considering information present in the baseline image.

By estimating atrophy values from a small number of variables, the current framework can be used as a simple simulator of disease progression where one can easily intervene on these inputs, including disease status, by simply changing the class. However, while the model can clearly differentiate between healthy and AD subjects, it is well documented that differentiation between MCI and AD is a complex task due to the heterogeneous nature of these disorders. This is reflected in the results from Fig. 4, and in particular for larger time-windows between scans, for which there are fewer data for the AD class. In the future, and in addition to exploring the use of imaging data as input to the atrophy estimator, we aim to include other metrics of cognitive assessment as input to the network, such as the Mini-Mental State Examination (MMSE) and Alzheimer's Disease Assessment Scale-Cognition (ADAS-Cog) in order to more accurately predict disease trajectories. We also aim to evaluate the impact of class imbalance on the network training, and address this by including more data from healthy subjects, as well as exploring techniques of oversampling and weighted loss when training the network.

In this work, we estimate region-wise atrophy maps, which are then used as input to the biomechanical model. In future work, and in order to model patient-specific trajectories of disease, we plan on using the region-wise atrophy estimates as priors to further compute subject-specific voxel-wise patterns that more accurately represent true atrophy patterns.

Note, although we focus on modelling brain atrophy with age, this proposed model can be translated to other tasks, including brain growth, and can support the use of different biomechanical models of tissue deformation.

Acknowledgments. The data used in this work was funded by the Alzheimer's Disease Neuroimaging Initiative (ADNI) (National Institutes of Health Grant U01 AG024904) and DOD ADNI (Department of Defense award number W81XWH-12-2-0012).

References

1. Bae, J.B., et al.: Identification of Alzheimer's disease using a convolutional neural network model based on T1-weighted magnetic resonance imaging. Sci. Rep. **10**(1), 1–10 (2020)
2. Balakrishnan, G., Zhao, A., Sabuncu, M.R., Guttag, J., Dalca, A.V.: VoxelMorph: a learning framework for deformable medical image registration. IEEE Trans. Med. Imaging 38(8), 1788–1800, August 2019. https://doi.org/10.1109/TMI.2019.2897538, http://arxiv.org/abs/1809.05231
3. Bass, C., et al.: Image synthesis with a convolutional capsule generative adversarial network, December 2018. https://openreview.net/forum?id=rJen0zC1lE
4. Bass, C., da Silva, M., Sudre, C., Tudosiu, P.D., Smith, S., Robinson, E.: ICAM: interpretable classification via disentangled representations and feature attribution mapping. In: Advances in Neural Information Processing Systems, vol. 33 (2020)
5. Baumgartner, C.F., Koch, L.M., Tezcan, K.C., Ang, J.X., Konukoglu, E.: Visual feature attribution using wasserstein GANs, June 2018. http://arxiv.org/abs/1711.08998
6. Cardoso, M.J., et al.: Geodesic information flows: spatially-variant graphs and their application to segmentation and fusion. IEEE Trans. Med. Imaging **34**(9), 1976–1988 (2015). https://doi.org/10.1109/TMI.2015.2418298
7. Carmichael, O., McLaren, D.G., Tommet, D., Mungas, D., Jones, R.N., Initiative, A.D.N., et al.: Coevolution of brain structures in amnestic mild cognitive impairment. NeuroImage **66**, 449–456 (2013)
8. Ferreira, D., et al.: Distinct subtypes of Alzheimer's disease based on patterns of brain atrophy: longitudinal trajectories and clinical applications. Sci. Rep. **7**(1), 1–13 (2017). https://doi.org/10.1038/srep46263
9. Frisoni, G.B., Fox, N.C., Jack, C.R., Scheltens, P., Thompson, P.M.: The clinical use of structural MRI in Alzheimer disease. Nat. Rev. Neurol. **6**(2), 67–77 (2010)
10. Jenkinson, M., Smith, S.: A global optimisation method for robust affine registration of brain images. Med. Image Anal. **5**(2), 143–156 (2001)
11. Khan, N.M., Abraham, N., Hon, M.: Transfer learning with intelligent training data selection for prediction of Alzheimer's disease. IEEE Access **7**, 72726–72735 (2019)

12. Khanal, B., Lorenzi, M., Ayache, N., Pennec, X.: A Biophysical Model of Shape Changes due to Atrophy in the Brain with Alzheimer's Disease. In: Golland, P., Hata, N., Barillot, C., Hornegger, J., Howe, R. (eds.) MICCAI 2014. LNCS, vol. 8674, pp. 41–48. Springer, Cham (2014). https://doi.org/10.1007/978-3-319-10470-6_6

13. Khanal, B., Lorenzi, M., Ayache, N., Pennec, X.: A biophysical model of brain deformation to simulate and analyze longitudinal MRIs of patients with Alzheimer's disease. NeuroImage **134**, 35–52 (2016). https://doi.org/10.1016/j.neuroimage.2016.03.061, http://www.sciencedirect.com/science/article/pii/S1053811916300052

14. Bigolin Lanfredi, R., Schroeder, J.D., Vachet, C., Tasdizen, T.: Interpretation of Disease Evidence for Medical Images Using Adversarial Deformation Fields. In: Martel, A.L., Abolmaesumi, P., Stoyanov, D., Mateus, D., Zuluaga, M.A., Zhou, S.K., Racoceanu, D., Joskowicz, L. (eds.) MICCAI 2020. LNCS, vol. 12262, pp. 738–748. Springer, Cham (2020). https://doi.org/10.1007/978-3-030-59713-9_71

15. Li, H., Habes, M., Wolk, D.A., Fan, Y.: Alzheimer's disease neuroimaging initiative and the australian imaging biomarkers and lifestyle study of aging: a deep learning model for early prediction of Alzheimer's disease dementia based on hippocampal magnetic resonance imaging data. Alzheimer's Dement. **15**(8), 1059–1070 (2019). https://doi.org/10.1016/j.jalz.2019.02.007, https://alz-journals.onlinelibrary.wiley.com/doi/abs/10.1016/j.jalz.2019.02.007

16. Pawlowski, N., Castro, D.C., Glocker, B.: Deep structural causal models for tractable counterfactual inference, June 2020. arXiv:2006.06485 [cs, stat], http://arxiv.org/abs/2006.06485

17. Rabinovici, G., et al.: Distinct MRI atrophy patterns in autopsy-proven Alzheimer's disease and frontotemporal lobar degeneration. Am. J. Alzheimer's Dis. Dement. & #x00AE; 22(6), 474–488 (2008)

18. Richman, D.P., Stewart, R.M., Hutchinson, J.W., Caviness, V.S.: Mechanical model of brain convolutional development. Sci. (New York, N.Y.) 189(4196), 18–21 (1975). doi: 10.1126/science.1135626

19. Rodriguez, E.K., Hoger, A., McCulloch, A.D.: Stress-dependent finite growth in soft elastic tissues. J. Biomech. **27**(4), 455–467 (1994). https://doi.org/10.1016/0021-9290(94)90021-3

20. Tallinen, T., Chung, J.Y., Biggins, J.S., Mahadevan, L.: Gyrification from constrained cortical expansion. Proc. Natl. Acad. Sci. **111**(35), 12667–12672 (2014). https://doi.org/10.1073/pnas.1406015111, https://www.pnas.org/content/111/35/12667

21. Tallinen, T., Chung, J.Y., Rousseau, F., Girard, N., Lefèvre, J., Mahadevan, L.: On the growth and form of cortical convolutions. Nat. Phys. **12**(6), 588–593 (2016). https://doi.org/10.1038/nphys3632, https://www.nature.com/articles/nphys3632

22. Xu, G., Knutsen, A.K., Dikranian, K., Kroenke, C.D., Bayly, P.V., Taber, L.A.: Axons pull on the brain, but tension does not drive cortical folding. J. Biomech. Eng. **132**(7), 071013 (2010). https://doi.org/10.1115/1.4001683

23. Young, J.M., Yao, J., Ramasubramanian, A., Taber, L.A., Perucchio, R.: Automatic generation of user material subroutines for biomechanical growth analysis. J. Biomech. Eng. **132**(10), 104505 (2010). https://doi.org/10.1115/1.4002375, https://www.ncbi.nlm.nih.gov/pmc/articles/PMC2996139/

Towards Self-explainable Classifiers and Regressors in Neuroimaging with Normalizing Flows

Matthias Wilms[1,2,3](✉), Pauline Mouches[1,2,3], Jordan J. Bannister[1,2,3],
Deepthi Rajashekar[1,2,3], Sönke Langner[4], and Nils D. Forkert[1,2,3]

[1] Department of Radiology, University of Calgary, Calgary, Canada
matthias.wilms@ucalgary.ca
[2] Hotchkiss Brain Institute, University of Calgary, Calgary, Canada
[3] Alberta Children's Hospital Research Institute, University of Calgary,
Calgary, Canada
[4] Institute for Diagnostic and Interventional Radiology, Pediatric and
Neuroradiology, University Medical Center Rostock, Rostock, Germany

Abstract. Deep learning-based regression and classification models are
used in most subareas of neuroimaging because of their accuracy and
flexibility. While such models achieve state-of-the-art results in many
different applications scenarios, their decision-making process is usually
difficult to explain. This black box behaviour is problematic when non-
technical users like clinicians and patients need to trust them and make
decisions based on their results. In this work, we propose to build self-
explainable generative classifiers and regressors using a flexible and efficient normalizing flow framework. We directly exploit the invertibility
of those normalizing flows to explain the decision-making process in a
highly accessible way via consistent and spatially smooth attribution
maps and counterfactual images for alternate prediction results. The
evaluation using more than 5000 3D MR images highlights the explainability capabilities of the proposed models and shows that they achieve
a similar level of accuracy as standard convolutional neural networks for
image-based brain age regression and brain sex classification tasks.

Keywords: Normalizing flows · Explainable AI

1 Introduction

Over the last decade, deep neural networks (DNNs) have revolutionized medical
image analysis in general and many areas of neuroimaging in particular [32].
Their accuracy is usually a result of the availability of large training data sets
that help the models to learn complex functions to map the input images to
the desired outputs. While this data-driven approach helps to learn accurate
mappings, it is also a major reason why DNNs are often deemed black boxes
that are difficult to analyze and interpret [32]. This lack of interpretability and

© Springer Nature Switzerland AG 2021
A. Abdulkadir et al. (Eds.): MLCN 2021, LNCS 13001, pp. 23–33, 2021.
https://doi.org/10.1007/978-3-030-87586-2_3

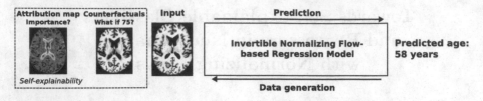

Fig. 1. Graphical overview of the proposed use of an invertible regression model for explainable brain age prediction. The invertible model can predict the age of a given input image (left to right) but can also be used to generate data (right to left) that helps to explain its decision-making process through interpretable voxel-level attribution maps and counterfactual images for alternate prediction results.

explainability is a major problem in neuroimaging applications where their success relies on the acceptance by non-technical users like clinicians and patients who need to trust the model-generated results [10,20]. Recently, considerable progress has been made to open up the so-called black boxes by developing mechanisms that help to explain and interpret the decisions made by DNN models [6,22]. Many strategies used for image-based DNNs like convolutional neural networks (CNNs) can be categorized as model-agnostic, post-hoc feature attribution methods [3,6,20,22], which query a trained model to determine what features contribute most to a specific decision. Popular representatives of this category are gradient-based approaches that generate saliency or attribution maps by computing the gradient of the decision function with respect to input features [23,24,27]. While those methods are easy to apply and popular in neuroimaging (e.g., [5,17]), they lack an easily accessible explanation of a DNN's decision as they mainly answer questions such as *"Which voxels affect the prediction most?"*. It can be argued that non-technical users like clinicians are more interested in questions that reveal the learned concepts [7]: *"Why was this image classified as A and not B?"* and *"How would this image look like if it was from a different class?"*. Therefore, approaches that analyze a given model's behaviour with respect to concepts have been proposed [8,14]. Those models provide a better comprehension of a DNN's decisions but are usually not able to generate meaningful alternate versions of the real images (counterfactuals) that would have resulted in different decisions and that are known to help in visually assessing the decision-making process [7,13]. More sophisticated approaches generate images for alternate decisions using a generative model such as a Generative Adversarial Network (GAN) that approximately inverts the DNN's decision-making process [19,25].

Instead of training an additional generative model for explanation purposes, it is more natural and consistent to build a self-explainable model [22]. This can, for example, be done by using invertible DNN architectures that not only solve the classification/regression task but are also able to systematically manipulate the inputs to explain their decisions via counterfactual images. Recently, generative classifiers based on normalizing flows (NFs, [15]) have gained popularity in computer vision [2,18,26]. Aside from notable exceptions like [11,31], NFs have rarely been used in the medical domain due to their high computational costs when applied to 3D data. This issue has recently been addressed in [30] where a deformation-based NF brain aging model was proposed that circumvents computational limitations by equipping the NF with a theoretically sound dimensionality reduction step. Although the model in [30] performs brain age regression, the authors do not exploit its invertibility for explainability purposes.

In this paper, we (1) propose to build invertible, self-explainable generative classifiers and regressors derived from the efficient NF approach of [30] and (2) describe how to directly exploit the generative/invertibility properties of the models to explain their decisions through more natural attribution maps as well as counterfactuals (see Fig. 1). To our knowledge, this paper is the first that proposes invertible generative classifiers and regressors using NFs for explainable AI for 3D neuroimaging data and which also shows that they are able to achieve results comparable to standard black box CNNs. Our approach relies on the architecture proposed in [30], but we directly apply it to 3D images instead of deformation fields and not only investigate its use for brain age regression but also for sex classification. We also draw from some explainability definitions and concepts proposed in [26], but our setup significantly differs from theirs.

2 Normalizing Flows as Generative Invertible Classifiers and Regressors

Our goal is to learn a NF-based decision function that maps a 3D input image $X : \mathbb{R}^3 \to \mathbb{R}$ to a scalar result $r \in \Omega$. For a continuous regression problem, $\Omega \equiv \mathbb{R}$ and for a binary classification problem, we define $\Omega \equiv [0, 1]$. Without loss of generality, we also define that $\mathbf{x} \in \mathbb{R}^{n_{\text{vox}}}$ is a sampled and vectorized version of X with n_{vox} being the number of image voxels. The decision function is then defined as $f : \mathbb{R}^{n_{\text{vox}}} \to \Omega \times \mathbb{R}^{n_{\text{vox}}-1}$ and maps an input vector of n_{vox} voxels to a vector of the same size with the first dimension representing the decision result. While such a definition seems counterintuitive at first, when using a NF, $f(\cdot)$ has to be a bijection to guarantee its invertibility. The $n_{\text{vox}} - 1$ dimensions of the output vector not encoding the decision result, contain the additional data needed to reconstruct the input $\mathbf{x}$ when applying the inverse $f^{-1}(\cdot)$ (see [30]).

The invertible function $f(\cdot)$ is then used in conjunction with the change-of-variable technique to define a conditional probabilistic generative model that allows to assign probabilities to images $\mathbf{x}$ given r by mapping the input space to a latent space on which simple priors can be imposed [30]:

$$p(\mathbf{x}|r) = p_Z(f_z(\mathbf{x})) p_R(f_r(\mathbf{x})|r) \left| \det\left(\frac{\partial f(\mathbf{x})}{\partial \mathbf{x}}\right) \right|. \quad (1)$$

Here, $p(\mathbf{x}|r)$ is the density of 3D images conditioned on a decision result r, $f_z : \mathbb{R}^{n_{\text{vox}}} \to Z$ is the part of $f(\cdot)$ that maps $\mathbf{x}$ to $Z \equiv \mathbb{R}^{n_{\text{vox}}-1}$, while $f_r : \mathbb{R}^{n_{\text{vox}}} \to R$ maps $\mathbf{x}$ to $R \equiv \Omega$. We use a factorized multivariate Gaussian prior for $\mathbf{z} \sim p_Z$, which covers the image variability independent of r, and assume that density $p_R(f_r(\mathbf{x})|r)$ is implicitly modeled by the regression or classification metric used (see Sect. 2.2). In summary, Eq. (1) defines a bidirectional model where $f_r(\cdot)$ solves the regression/classification problem (predictive direction), while $f^{-1}(\cdot)$ and sampling from the priors allows to generate data (generative direction).

2.1 Manifold-Constrained NFs for Efficient 3D Data Processing

The major challenge of the setup presented above is the high dimensionality of the inputs $\mathbf{x} \in \mathbb{R}^{n_{\text{vox}}}$ (millions of voxels for 3D images), which implies that enormous computational resources are required to learn $f(\cdot)$. We, therefore, follow the dimensionality reduction approach proposed in [30] for deformation fields and adapt it to gray value images. The basic idea is to approximate $f(\cdot)$ through a two-step process as $f(\mathbf{x}) \approx e(\mathbf{x}) = h \circ g(\mathbf{x})$. Here, $g : \mathbb{R}^{n_{\text{vox}}} \rightarrow \mathbb{R}^{n_{\text{dim}}}$ first projects $\mathbf{x}$ to a lower dimensional manifold of dimensionality n_{dim} before a standard NF-based bijection $h : \mathbb{R}^{n_{\text{dim}}} \rightarrow \mathbb{R}^{n_{\text{dim}}}$ maps this space to the model's latent space whose first dimension holds the regression/classification result and on which priors are imposed. While $g(\cdot)$ is not a bijection, we assume that it is a chart of the n_{dim}-dimensional manifold and invertible for manifold elements. Hence, $e : \mathbb{R}^{n_{\text{vox}}} \rightarrow \mathbb{R}^{n_{\text{dim}}}$ can be used as a replacement for $f(\cdot)$ in Eq. (1) if we assume that the information lost is not important for the task at hand.

Similar to what has been done in [30] for deformations, we assume that 3D gray value images can be sufficiently represented by a n_{dim}-dimensional affine subspace with translation vector $\overline{\mathbf{x}} \in \mathbb{R}^{n_{\text{vox}}}$ and a matrix $\mathbf{Q} \in \mathbb{R}^{n_{\text{vox}}} \times \mathbb{R}^{n_{\text{dim}}}$ composed of n_{dim} orthogonal columns. This subspace is estimated from training data in closed form via principal component analysis (PCA) and retaining the top n_{dim} eigenvectors. Then, $\mathbf{y} = g(\mathbf{x}) = \mathbf{Q}^+(\mathbf{x} - \overline{\mathbf{x}})$ with $\mathbf{Q}^+$ being the pseudoinverse of $\mathbf{Q}$ and $g(\mathbf{y})^{-1} = \mathbf{Q}\mathbf{y} + \overline{\mathbf{x}}$. Here, $[r, \mathbf{z}] = e(\mathbf{x}) = h \circ g(\mathbf{x})$ solves the regression/classification problem by mapping $\mathbf{x}$ to r (and $\mathbf{z} \in \mathbb{R}^{n_{\text{dim}}-1}$) and its inverse $\mathbf{x} = e^{-1}([r, \mathbf{z}]) = g^{-1} \circ h^{-1}([r, \mathbf{z}])$ can be utilized to generate new images.

2.2 Implementation Details and Model Training

Learning $e(\cdot)$ from data is equivalent to learning $h(\cdot)$ as $g(\cdot)$ can be pre-computed via PCA. For $[r, \mathbf{z}] = h(\mathbf{y}) = h_{n_{\text{lay}}} \circ \cdots \circ h_i \circ \cdots \circ h_1(\mathbf{y})$, we follow the typical NF paradigm and use a sequence of n_{lay} easily invertible sub-functions/layers $h_i : \mathbb{R}^{n_{\text{dim}}} \rightarrow \mathbb{R}^{n_{\text{dim}}}$. Each $\mathbf{b} = h_i(\mathbf{a})$ is an affine coupling layer that transforms its input vector $\mathbf{a} = [\mathbf{a}_1, \mathbf{a}_2]$ to $\mathbf{b} = [\mathbf{b}_1, \mathbf{b}_2]$ by splitting the inputs into two equally sized parts and applying a learnable affine transform (see also [15,30]):

$$\mathbf{b}_1 = \exp\big(s_i(\mathbf{a}_2)\big) \odot \mathbf{a}_1 + t_i(\mathbf{a}_2) \quad \text{and} \quad \mathbf{b}_2 = \mathbf{a}_2 . \tag{2}$$

Here, $\odot$ denotes an element-wise multiplication and $s_i(\cdot)$ and $t_i(\cdot)$ are fully-connected neural networks with n_{hid} hidden layers and ReLU activations. An affine coupling layer can be inverted by reversing the affine transformation without having to invert $s_i(\cdot)$ and $t_i(\cdot)$ [15,30]. To learn the weights of all $s_i(\cdot)/t_i(\cdot)$ based on a training set $\{(\mathbf{x}_j, r_j)\}_{j=1}^{n_{\text{sbj}}}$ of n_{sbj} subjects with images $\mathbf{x}_j$ and ground-truth targets r_j, we minimize the negative log-likelihood of Eq. (1) for $e(\cdot)$. Using a Gaussian prior for the difference between r_j and prediction $e_r(\mathbf{x}_j)$ and a unit-diagonal multivariate Gaussian prior for $\mathbf{z} \sim p_Z$ resulting from $e_z(\mathbf{x}_j)$, gives

$$\mathcal{L} = \frac{1}{n_{\text{sbj}}} \sum_{j=1}^{n_{\text{sbj}}} \left(\frac{1}{2} \left(\sigma^{-2} \|r_j - e_r(\mathbf{x}_j)\|_2^2 + \|e_z(\mathbf{x}_j))\|_2^2 \right) - \log\left|\det\left(\frac{\partial e(\mathbf{x})}{\partial \mathbf{x}}\right)\right| \right). \quad (3)$$

The parameter σ can be used to balance the target fit and the adherence of the additional unrelated variability to the imposed prior. For a regression problem, Eq. (3) minimizes the L^2 loss between prediction and ground-truth as in [30], while for a binary classification problem, we first pass $e_r(\mathbf{x}_j)$ through a sigmoid function to map the output to the unit interval before computing the difference.

3 Explainable AI with Normalizing Flows

The goal is now to exploit the invertibility of the learned decision function $[r, \mathbf{z}] = e(\mathbf{x}) = h \circ g(\mathbf{x})$ to explain the model's decision-making process in regression and classification scenarios in a highly accessible way through a derivative-based attribution map (Sect. 3.1) and counterfactual images (Sect. 3.2).

3.1 Derivative-Based Attribution Map of the Inverse

Standard gradient-based approaches to compute attribution maps for DNNs like Grad-CAM and SmoothGrad [23, 27] (see also Sect. 1), rely on the gradient of the decision function with respect to the input image or derived feature maps. The resulting attribution maps indicate voxel locations that influenced the decision made. However, we argue that, from an interpretability point of view, this is an unnatural way of analyzing the model and that a more natural approach is to explore what effect manipulations of the decision result would have on the input image $\mathbf{x}$ (see also [26] and Sect. 1). Such an analysis can be easily carried out with an invertible NF model by computing the partial derivative of the inverse $e^{-1}([r, \mathbf{z}])$ with respect to prediction result $r = e_r(\mathbf{x})$ (with $\mathbf{z} = e_z(\mathbf{x})$):

$$\frac{\partial}{\partial r} e^{-1}([r, \mathbf{z}]) = \frac{\partial}{\partial r} \left(\mathbf{Q} h^{-1}([r, \mathbf{z}]) + \bar{\mathbf{x}} \right) = \mathbf{Q} \frac{\partial}{\partial r} h^{-1}([r, \mathbf{z}]) . \quad (4)$$

The partial derivative $\frac{\partial}{\partial r} h^{-1}([r, \mathbf{z}])$ can be conveniently computed via automatic differentiation. We then use a normalized version as an attribution map that can be visualized in the image space. Utilizing the inverse of the decision function has the benefit that only voxels directly related to r will be highlighted while the remaining unrelated image information projected to $\mathbf{z}$ will have no effect.

3.2 Counterfactual Images for Systematic Analyses

Geometrically, the partial derivative in Eq. (4) defines the tangent to the curve parameterized by $e^{-1}([r, \mathbf{z}])$ for a fixed vector $\mathbf{z}$. Assuming that the training process described in Sect. 2.2 successfully disentangles r (= decision result) and $\mathbf{z}$

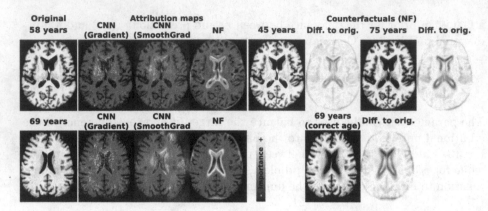

Fig. 2. Explainability results for the brain age regression task for two test subjects at the age of 58 years (top) and 69 years (bottom) when using the baseline CNN and the NF-based model. Both models correctly estimate the age of the first subject (58 years; rounded) while both fail to do so for the second one (CNN: 55 years; NF: 52 years). For the CNN model, vanilla gradients and SmoothGrad maps are visualized. For the NF-based model, partial derivatives of the inverse, counterfactual images, and associated difference images to the original image (blue: negative gray value diff., red: positive diff.; white: no diff.) are visualized. Bottom row counterfactual: The brain the model would have expected to see for the correct prediction (69 yrs. instead of 52 yrs.). (Color figure online)

(= unrelated information) for a given input image $\mathbf{x}$, we can follow the curve to generate meaningful artificial images for alternate decision results $r + \delta r$ to systematically analyze the model's decision-making process and the learned concept (*"How would this image look like if r was different?"*; see Sect. 1). Those artificial images represent alternate realities and are usually called counterfactuals [26]. Given prediction $[r, \mathbf{z}] = e(\mathbf{x})$ for an input image $\mathbf{x}$, we define a counterfactual image $\mathbf{x}_{\delta r}$ for a modified decision result $r + \delta r$ and fixed $\mathbf{z}$ as

$$\mathbf{x}_{\delta r} = e^{-1}([r + \delta r, \mathbf{z}]) + \mathbf{n}_{\mathbf{x}} . \tag{5}$$

Vector $\mathbf{n}_{\mathbf{x}} \in \mathbb{R}^{n_{\text{vox}}}$ is the information of $\mathbf{x}$ being lost when projecting the image to the affine subspace; $\mathbf{n}_{\mathbf{x}} = (g^{-1} \circ g(\mathbf{x})) - \mathbf{x}$. We add this residual information to the generated counterfactual image to improve its visual appearance by substantially reducing the blurriness. Keeping $\mathbf{z}$ fixed allows us to generate images $\mathbf{x}_{\delta r}$ highly similar to $\mathbf{x}$ that would lead to another decision result $r + \delta r$ when applying $e(\cdot)$. We, therefore, argue that a NF-based model is inherently self-explainable.

4   Experiments and Results

The evaluation aims at showing that NF-based generative classifiers and regressors (1) achieve competitive results for typical classification/regression problems in neuroimaging and (2) are better at explaining their decisions when compared to a standard black box CNN model. The tasks being analyzed here are brain age regression and brain-based sex classification using structural T1 MR images.

Data: We use T1-weighted brain MR images of 5287 healthy adults from five data bases (SHIP [28]: 3164 subjects; IXI[1]: 563 subjects; SALD [29]: 494 subjects; DLBS[2]: 309 subjects; OASIS-3 [16]: 757 subjects). For all subjects, age data (age range: 20–90 years) as well as their sex is available (females: 55%). All 5287 images are first pre-processed (N4 bias·correction and skull-stripping [12]), affinely registered and histogram matched to the SRI24 atlas [21] (cropped to $173{\times}211{\times}155$ voxels; isotropic 1 mm spacing), and finally split into independent training (4281 subjects), test (684 subjects), and validation subsets (322 subjects) via age- and sex-stratified random sampling.

Experimental Design: Based on the 4281 training subjects, independent NF-based models as described in Sect. 2 are build for both tasks. Brain age regression is a continuous regression problem (true age vs. predicted age) and sex classification is a binary classification scenario (true sex vs. predicted sex). For both problems/models, the same architecture is used: affine subspace with $n_{\mathrm{dim}} = 500$, $n_{\mathrm{lay}} = 16$ affine coupling layers, fully-connected scaling/translation networks with $n_{\mathrm{hid}} = 2$ hidden layers of width 32. During training, Eq. (3) is minimized for 20k epochs with an AdamW optimizer, a learning rate of 10^{-4}, and $\sigma = 0.16$ (age)/$\sigma = 0.1$ (sex). Parameters were chosen based on experiments on the validation data; see also [30]. Training takes 3 h on a NVIDIA Quadro P4000 GPU with a TensorFlow 2.2 implementation. As a baseline, we also train a classical CNN [4] frequently used in neuroimaging for both tasks.

Results: Our NF-based generative brain age regression model achieves a mean absolute error (MAE) between true age and predicted age of $4.83{\pm}3.60$ years for all test subjects, while the MAE for the baseline CNN model from [4] is $4.45{\pm}3.33$ years. For the brain sex classification task, our NF-based classifier achieves an accuracy of 90.10% when using a 0.5 threshold after the sigmoid function and an area under the curve of the receiver operating characteristic (AUROC) of 0.97. The accuracy of the baseline CNN model is 92.98% (0.5 threshold) and the AUROC is 0.97. While the MAE and accuracy values are slightly better for the baseline CNN, all differences are not statistically significant (paired Wilcoxon signed rank test; alpha level: 0.05; age: $p = 0.09$; sex: $p = 0.49$).

Figures 2 (regression) and 3 (classification) show explainable AI results for both tasks and models. For the NF-based models, derivative-based attribution maps of the inverse (see Sect. 3.1) and counterfactual images for alternate prediction results (see Sect. 3.2) are visualized. In addition, vanilla gradients as well as SmoothGrad ([27], regression only, parameters from [17]) and Grad-CAM attribution maps ([23], classification only) are shown for the baseline CNN. For both tasks and in comparison to the other attribution maps, it is immediately obvious that the maps of the NF-based models are often more consistent and/or spatially smoother. We argue that those properties make it easier to interpret the maps

---

[1] https://brain-development.org/ixi-dataset/.
[2] http://fcon_1000.projects.nitrc.org/indi/retro/dlbs.html.

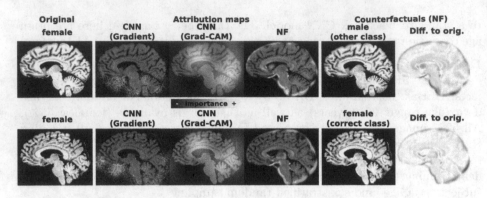

Fig. 3. Explainability results for the sex classification task for two female test subjects. The CNN model correctly classifies both subjects, while the NF-based model classifies the bottom one as being male (see also counterfactual image). For the CNN model, vanilla gradients and Grad-CAM maps are visualized. See also caption of Fig. 2.

while also increasing their trustworthiness. It also highlights that the NF-based models implicitly learn a consistent concept of the task at hand. It needs to be highlighted in particular that the concept learned for the brain age regression problem is in agreement with what is known about the normal aging process [9] as the model focuses on areas where aging-related atrophy is usually most visible (ventricles and gyri/sulci). The counterfactual images generated for the age regression task also help to understand the concept learned and are especially useful when analyzing why the NF model severely underestimated the age of the second subject (52 years instead of 69 years). From the counterfactual image for the correct age (69 years), it can be seen that the model would have expected to see larger ventricles, indicating a more advanced state of atrophy.

The sex classification results are harder to interpret/verify as the average global volume difference between male and female brains cannot be used by the models as this information was removed by the affine registration during pre-processing. Interestingly, the NF-based model still uses some volume information (brains of males show more atrophy) as highlighted by the attribution maps and the difference images. This may indicate that a bias exists in the training data or was introduced during pre-processing and which would be harder to detect when using CNN-related maps. The NF model additionally believes that females have a larger cerebellum (see counterfactual for second subject in Fig. 3) than males, which is in agreement with the pediatric study results in [1].

5   Conclusion

In this paper, we proposed to build self-explainable generative classifiers and regressors based on invertible NFs that are easy to train and directly applicable to 3D neuroimaging data. Our evaluation showed that the proposed models achieve competitive results when compared to a standard CNN model for brain age regression and brain sex classification. Because of their invertibility, the resulting models can be utilized to generate smooth and consistent attribution maps that directly visualize the concepts they are using. Furthermore, they can generate realistic counterfactual images for alternate prediction results that help to systematically analyze a model's behaviour in a highly accessible way. All of this is possible without using any additional models or methods. The models proposed in this work are also fully-functional probabilistic generative models. This property has not been explicitly exploited here, but it allows, for example, to systematically sample data or to generate conditional templates (see [30] for examples). Future work will focus on additional regression and classification tasks in neuroimaging and the addition of other baseline methods for explainable AI to the evaluation. In summary, we see this work as an important step towards normalizing flow-based, self-explainable models in neuroimaging and believe that in the future improvements with respect to their accuracy can be expected by incorporating recent advances from the machine learning community [2,18].

Acknowledgements. This work was supported by a T. Chen Fong postdoctoral fellowship and the River Fund at Calgary Foundation.

References

1. Adeli, E., et al.: Deep learning identifies morphological determinants of sex differences in the pre-adolescent brain. Neuroimage **223**, 117293 (2020)
2. Ardizzone, L., Mackowiak, R., Rother, C., Köthe, U.: Training normalizing flows with the information bottleneck for competitive generative classification. NeurIPS **33** (2020)
3. Arrieta, A.B., et al.: Explainable artificial intelligence (XAI): concepts, taxonomies, opportunities and challenges toward responsible AI. Inf. Fusion **58**, 82–115 (2020)
4. Cole, J.H., et al.: Predicting brain age with deep learning from raw imaging data results in a reliable and heritable biomarker. Neuroimage **163**, 115–124 (2017)
5. Eitel, F., Ritter, K.: Testing the robustness of attribution methods for convolutional neural networks in MRI-based alzheimer's disease classification. In: Suzuki, K., et al. (eds.) ML-CDS/IMIMIC -2019. LNCS, vol. 11797, pp. 3–11. Springer, Cham (2019). https://doi.org/10.1007/978-3-030-33850-3_1
6. Gilpin, L.H., Bau, D., Yuan, B.Z., Bajwa, A., Specter, M., Kagal, L.: Explaining explanations: an overview of interpretability of machine learning. In: 2018 IEEE 5th International Conference on Data Science and Advanced Analytics (DSAA), pp. 80–89. IEEE (2018)
7. Goyal, Y., Wu, Z., Ernst, J., Batra, D., Parikh, D., Lee, S.: Counterfactual visual explanations. In: ICML, pp. 2376–2384 (2019)
8. Graziani, M., Andrearczyk, V., Marchand-Maillet, S., Müller, H.: Concept attribution: explaining CNN decisions to physicians. Comput. Biol. Med. **123**, 103865 (2020)
9. Hedman, A.M., van Haren, N.E., Schnack, H.G., Kahn, R.S., Hulshoff Pol, H.E.: Human brain changes across the life span: a review of 56 longitudinal magnetic resonance imaging studies. Human Brain Mapp. **33**(8), 1987–2002 (2012)
10. Holzinger, A., Biemann, C., Pattichis, C.S., Kell, D.B.: What do we need to build explainable AI systems for the medical domain? (2017) arXiv:1712.09923

11. Hwang, S.J., Tao, Z., Kim, W.H., Singh, V.: Conditional recurrent flow: conditional generation of longitudinal samples with applications to neuroimaging. In: CVPR, pp. 10692–10701 (2019)
12. Isensee, F., et al.: Automated brain extraction of multisequence MRI using artificial neural networks. Human Brain Mapp. **40**(17), 4952–4964 (2019)
13. Jeyakumar, J.V., Noor, J., Cheng, Y.H., Garcia, L., Srivastava, M.: How can i explain this to you? an empirical study of deep neural network explanation methods. NeurIPS **33** (2020)
14. Kim, B., Wattenberg, M., Gilmer, J., Cai, C., Wexler, J., Viegas, F., et al.: Interpretability beyond feature attribution: quantitative testing with concept activation vectors (TCAV). In: ICML, pp. 2668–2677. PMLR (2018)
15. Kobyzev, I., Prince, S., Brubaker, M.: Normalizing flows: an introduction and review of current methods. IEEE TPAMI, 1–1 (2020)
16. LaMontagne, P.J., et al.: Oasis-3: longitudinal neuroimaging, clinical, and cognitive dataset for normal aging and alzheimer disease. medRxiv (2019)
17. Levakov, G., Rosenthal, G., Shelef, I., Raviv, T.R., Avidan, G.: From a deep learning model back to the brain–identifying regional predictors and their relation to aging. Human Brain Mapp. **41**(12), 3235–3252 (2020)
18. Mackowiak, R., Ardizzone, L., Köthe, U., Rother, C.: Generative classifiers as a basis for trustworthy computer vision. arXiv:2007.15036 (2020)
19. Narayanaswamy, A., et al.: Scientific discovery by generating counterfactuals using image translation. In: Martel, A.L., et al. (eds.) MICCAI 2020. LNCS, vol. 12261, pp. 273–283. Springer, Cham (2020). https://doi.org/10.1007/978-3-030-59710-8_27
20. Reyes, M., et al.: On the interpretability of artificial intelligence in radiology: challenges and opportunities. Radiol. Artif. Intell. **2**(3), e190043 (2020)
21. Rohlfing, T., Zahr, N.M., Sullivan, E.V., Pfefferbaum, A.: The SRI24 multichannel atlas of normal adult human brain structure. Human Brain Mapp. **31**(5), 798–819 (2010)
22. Samek, W., Montavon, G., Lapuschkin, S., Anders, C.J., Müller, K.R.: Explaining deep neural networks and beyond: a review of methods and applications. Proc. IEEE **109**(3), 247–278 (2021)
23. Selvaraju, R.R., Cogswell, M., Das, A., Vedantam, R., Parikh, D., Batra, D.: Gradcam: visual explanations from deep networks via gradient-based localization. In: CVPR, pp. 618–626 (2017)
24. Simonyan, K., Vedaldi, A., Zisserman, A.: Deep inside convolutional networks: visualising image classification models and saliency maps. arXiv:1312.6034 (2013)
25. Singla, S., Pollack, B., Wallace, S., Batmanghelich, K.: Explaining the black-box smoothly-a counterfactual approach. arXiv:2101.04230 (2021)
26. Sixt, L., Schuessler, M., Weiß, P., Landgraf, T.: Interpretability through invertibility: a deep convolutional network with ideal counterfactuals and isosurfaces (2021). https://openreview.net/forum?id=8YFhXYe1Ps
27. Smilkov, D., Thorat, N., Kim, B., Viégas, F., Wattenberg, M.: Smoothgrad: removing noise by adding noise. arXiv:1706.03825 (2017)
28. Völzke, H., et al.: Cohort profile: the study of health in pomerania. Int. J. Epidemiol. **40**(2), 294–307 (2011)
29. Wei, D., Zhuang, K., Chen, Q., Yang, W., Liu, W., Wang, K., Sun, J., Qiu, J.: Structural and functional MRI from a cross-sectional southwest university adult lifespan dataset (sald). BioRxiv, p. 177279 (2017)

30. Wilms, M., et al.: Bidirectional Modeling and Analysis of Brain Aging with Normalizing Flows. In: Kia, S.M., et al. (eds.) MLCN/RNO-AI -2020. LNCS, vol. 12449, pp. 23–33. Springer, Cham (2020). https://doi.org/10.1007/978-3-030-66843-3_3
31. Zhen, X., Chakraborty, R., Yang, L., Singh, V.: Flow-based generative models for learning manifold to manifold mappings. arXiv:2012.10013 (2020)
32. Zhou, S.K., et al.: A review of deep learning in medical imaging: image traits, technology trends, case studies with progress highlights, and future promises. arXiv:2008.09104 (2020)

Patch vs. Global Image-Based Unsupervised Anomaly Detection in MR Brain Scans of Early Parkinsonian Patients

Verónica Muñoz-Ramírez[1,2], Nicolas Pinon[3], Florence Forbes[2], Carole Lartizen[3], and Michel Dojat[1(✉)]

[1] Univ. Grenoble Alpes, Inserm U1216, CHU Grenoble Alpes, Grenoble Institut des Neurosciences, 38000 Grenoble, France
{veronica.munoz-ramirez,michel.dojat}@univ-grenoble-alpes.fr
[2] Univ. Grenoble Alpes, Inria, CNRS, Grenoble INP, LJK, 38000 Grenoble, France
florence.forbes@inria.fr
[3] Univ. Lyon, CNRS, Inserm, INSA Lyon, UCBL, CREATIS, UMR5220, U1206, 69621 Villeurbanne, France
{nicolas.pinon,carole.lartizen}@creatis.insa-lyon.fr

Abstract. Although neural networks have proven very successful in a number of medical image analysis applications, their use remains difficult when targeting subtle tasks such as the identification of barely visible brain lesions, especially given the lack of annotated datasets. Good candidate approaches are patch-based unsupervised pipelines which have both the advantage to increase the number of input data and to capture local and fine anomaly patterns distributed in the image, while potential inconveniences are the loss of global structural information. We illustrate this trade-off on Parkinson's disease (PD) anomaly detection comparing the performance of two anomaly detection models based on a spatial auto-encoder (AE) and an adaptation of a patch-fed siamese auto-encoder (SAE). On average, the SAE model performs better, showing that patches may indeed be advantageous.

Keywords: Parkinson's disease · Anomaly detection · Patches · Siamese networks · Auto-encoders

1 Introduction

Medical imaging represents the largest percentage of data produced in healthcare and thus a particular interest has emerged in deep learning (DL) methods

Data used in the preparation of this article were obtained from the Parkinson‚s Progression Markers Initiative (PPMI) database (www.ppmi-info.org/data).
VMR is supported by a grant from NeuroCoG IDEX-UGA (ANR-15-IDEX-02). This work is partially supported by the French program "Investissement d'Avenir" run by the Agence Nationale pour la Recherche (ANR-11-INBS-0006).
VMR and NP contributed equally to this work.

A. Abdulkadir et al. (Eds.): MLCN 2021, LNCS 13001, pp. 34–43, 2021.
https://doi.org/10.1007/978-3-030-87586-2_4

to create support tools for radiologists to analyze multimodal medical images, segment lesions and detect subtle pathological changes that even an expert eye can miss. The vast majority of these methods are based on supervised models which require to be trained on large series of annotated data, time- and resource-consuming to generate.

Over the years, several publicly available neuroimaging databases have been curated and completed with annotations. Some of the most prominent ones are: MSSEG, for multiple sclerosis lesion segmentation [4]; BRATS, for brain tumor segmentation [14]; ISLES, for ischemic stroke lesion segmentation [13]; and mTOP for mild traumatic injury outcome prediction [12]. Challenges, namely those of MICCAI, are organized regularly to showcase the latest technological advancements and push the community towards better performances. However, there are several neurological diseases seldom studied due to the small size and subtlety of the lesions they present. This is the case of vascular disease, epilepsy, and most neurodegenerative diseases in their early stages. The main challenge for such pathologies indeed, is to identify the variability of the pathological patterns on images where the lesion is barely seen or not visible.

Unsupervised methods are good candidates to tackle both the lack of anno-tated examples and the subtlety of brain scan anomalies [3,19]. They rely on networks that learn to encode normal brain patterns in such a manner that any atypical occurrence can be identified by the inability of the network to reproduce it. Auto-encoders (AE), variational auto-encoders (VAE) [10] and generative adversarial networks (GAN) [7], have been extensively used as building blocks for unsupervised anomaly detection due to their ability to learn high-dimensional distributions [3].

Parkinson's Disease (PD) is a neurodegenerative disorder that is only iden-tifiable through routine MR scans at an advanced stage. Nevertheless, the man-ifestation of non-motor symptoms, years before the apparition of the first motor disturbances, suggests the presence of physio-pathological differences that could allow for earlier PD diagnosis. PD afflicts patients for as many as one to two decades of their lives and current treatments can only attenuate some motor manifestations [21]. Therefore, reducing the gap between diagnosis and the onset of the neurodegenerative process is of paramount importance to identify person-alized treatments that would significantly slow its natural progression. Unsuper-vised anomaly detection models are here employed to explore such challenging MR data analysis.

In a previous work [16], we compared deterministic and variational, spatial and dense autoencoders for the detection of subtle anomalies in the diffusion parametric maps of *de novo* (i.e., newly diagnosed and without dopaminergic treatment) PD patients from the PPMI database [11]. Our results, while prelim-inary, offered compelling evidence that DL models are useful to identify subtle anomalies in early PD, even when trained with a moderate number of images and only two parametric maps as input.

Our goal in this paper is to compare an improved anomaly detection pipeline based on a deterministic spatial auto-encoder, hereafter simply referred as AE, to an adaptation of patch-based siamese auto-encoder (SAE) proposed in [1]. This architecture was originally intended to the detection of subtle epileptic lesions, application for which it achieved promising results.

One important difference between the two compared architectures is the dimension of the input and output data. While AE were trained on 2D transverse slices, thus capturing a global pattern in the image, SAE were trained on small patches sampled throughout the data, making them more suitable to capture fine patterns but losing global structural integrity. Through this comparison we aim to analyze the advantages of patch-fed architectures for the identification of subtle and local abnormalities as well as to propose an alternative for anomaly detection in moderate size image datasets.

2 Brain Anomaly Detection Pipeline

The anomaly detection task with auto-encoders can be formally posed as follows:

- An auto-encoder is first trained to reconstruct normal samples as accurately as possible. This network is composed of two parts: an encoder (1) that maps the input data into a lower dimensional latent space, assumed to contain important image features, and a decoder (2) that maps the code from the latent space into an output image.
- When fed by an unseen image, this trained network produces a reconstructed image from its sampled latent distribution which is the counterpart 'normal' part of the input image.
- *Reconstruction error* maps, computed as the difference between the input and output images, are thus assumed to highlight anomalous regions of the input data.
- *Anomaly scores* at the voxel, region of interest or image levels may then be derived from the post-processing of these *reconstruction error* maps.

In this work, we present a general framework for unsupervised brain anomaly detection based on auto-encoders to produce reconstruction error maps and a novel post-processing step to derive per-region anomaly scores.

2.1 Autoencoder Architectures

We constructed and evaluated two auto-encoder models: a classic auto-encoder (AE) and a siamese auto-encoder (SAE). Both models are fully-convolutional. Their architectures are displayed in Fig. 1 and their differences are detailed below.

Classic Auto-Encoder: This architecture consists of 5 convolutional layers that go from input to bottleneck and 5 transposed convolutional layers going from bottleneck to output. The output of the encoder network is directly the latent vector z and the loss function was simply the L_1-norm reconstruction error between input $\mathbf{x}$ and output $\hat{\mathbf{x}}$:

$$L_{AE}(\mathbf{x}) = \|\mathbf{x} - \hat{\mathbf{x}}\|_1 \tag{1}$$

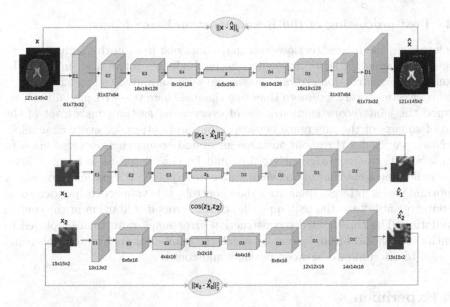

Fig. 1. Classical auto-encoder (AE) on top, Siamese auto-encoder (SAE) at the bottom

Siamese Auto-Encoder: The siamese autoencoder (SAE) model [1] consists of two identical convolutional autoencoders with shared parameters. The SAE receives a pair of patches $(\mathbf{x_1}, \mathbf{x_2})$ at input that are propagated through the network, yielding representations $\mathbf{z_t} \in \mathcal{Z}, t = (1,2)$ in the middle layer bottleneck. The second term of the loss function L_{SAE} (Eq. 2) is designed to maximize the cosine similarity between $\mathbf{z_1}$ and $\mathbf{z_2}$. This constraints patches that are "similar" to be aligned in the latent space. Unlike standard siamese architectures where *similar* and *dissimilar* pairs are presented to the network, Alaverdyan et al. [1] proposed to train this architecture on *similar* pairs only and compensate for the lack of *dissimilar* pairs through a regularizing term that prevents driving the loss function to 0 by mapping all patches to a constant value. This term is defined as the mean squared error between the input patches and their reconstructed outputs. The proposed loss function for a single pair hence is:

$$L_{SAE}(\mathbf{x_1}, \mathbf{x_2}) = \sum_{t=1}^{2} ||\mathbf{x_t} - \hat{\mathbf{x}}_t||_2^2 - \alpha \cdot cos(\mathbf{z_1}, \mathbf{z_2}) \qquad (2)$$

where $\hat{\mathbf{x}}_t$ is the reconstructed output of the patch $\mathbf{x_t}$ while $\mathbf{z_t}$ is its representation in the middle layer bottleneck and α an hyperparameter that controls the trade-off between the two terms.

As depicted on Fig. 1, the encoder part is composed of 3 convolutional layers and one maxpooling layer in-between the first and second convolutions, while the (non symmetrical) decoder part is composed of 4 convolutional layers, with an upsampling layer in-between the second and third convolutional layer.

2.2   Post-processing of the Reconstruction Error Maps

We leveraged the reconstruction error maps obtained from both architectures to generate an anomaly score, following the methodology introduced in [15]. The voxel-wise reconstruction errors in one image were computed as $||x_i - \hat{x}_i||_1$. Since the architectures were fed more than one channel (here two MR modalities), we defined the joint reconstruction error of every voxel as the square root of the sum of squares of the difference between input and output for every channel.

Next, we fixed a threshold on these generated *reconstruction error maps* to decide whether or not a given voxel should be considered as abnormal, here-after called the *abnormality threshold*. Since we expected PD patients to exhibit abnormal voxels in larger quantities than controls, this value corresponded to an extreme quantile (e.g. the 98% quantile) of the errors distribution in the control population. The thresholded reconstruction error maps were then employed to identify anatomical brain regions for which the number of abnormal voxels could be used to discriminate between patients and controls.

3   Experiments

3.1   Data

The dataset used in this work consisted of DTI MR scans of 57 healthy controls and 129 *de novo* PD patients selected from the PPMI database. All images were acquired with the same MR scanner model (3T Siemens Trio Tim) and configured with the same acquisition parameters. Only one healthy control was taken out of the study due to important artifacts in the images.

From these images, two measures, mean diffusivity (MD) and fractional anisotropy (FA), were computed using MRtrix3.0. Values of FA and MD were normalized into the range $[0, 1]$. The images were spatially normalized to the standard brain template of the Montreal Neurological Institute (MNI) with a non-linear deformation. The resulting MD and FA parameter maps were of dimension $121 \times 145 \times 121$ with a voxel size of $1.5 \times 1.5 \times 1.5 \, \text{mm}^3$.

The control dataset was divided into 41 training controls and 15 testing controls to avoid data leakage. This division was effectuated in 10 different manners through a bootstrap procedure in order to assess the generalization of our predictions as advised in [17]. We took special care to maintain an age average around 61 years old for all the training and test population as well as a 40–60 proportion of females and males.

Once the models were trained with one of the 10 training sets, they were evaluated with the corresponding healthy control test set and the PD dataset (age: 62 y. ± 9; sex: 48 F).

3.2   Training of the Auto-Encoders

AE Models. The training dataset of the AE models consisted of 1640 images corresponding to 40 axial slices around the center of the brain for each of the

41 training control subjects. The AE models were trained for 160 epochs, with a learning rate of 10^{-3}. 3×3 kernels were convoluted using padding of 1 pixel and a stride of $(2, 2)$. The bottleneck dimensions were h=4, w=5 and c=256. There were no pooling layers. Implementation was done in Python 3.6.8, PyTorch 1.0.1, CUDA 10.0.130 and trained on a NVIDIA GeForce RTX 2080 Ti GPU with batches of 40 images. After each convolutional layer, batch normalization [8] was applied for its regularization properties. The rectified linear unit (ReLU) activation function was employed in each layer except the last, for which a sigmoïd was preferred. The loss functions were optimized using Adam [9].

SAE Models. The SAE model was trained with 600 000 patches of size $15\times15\times2$ ($\sim$15 000 patches per subject). The model was trained for 30 epochs, with a learning rate of 1×10^{-3}. Bottleneck dimensions were h=2, w=2 and c=16. Maxpooling and upsampling layers were used, as detailled before. No batch-normalization layers were used. Activation function for every convolutional layer was the rectified linear unit (ReLU) and the sigmoid function was used in the last layer. The kernel size and the numbers of filters were 3×3 and 16 respectively for all convolution blocks but the final one with 2×2 and 2 (equal to the number of channels) respectively. The stride was 1 for all convolution blocks. The maxpooling/upsampling factor was 2. Implementation was done in Python 3.8.10, Tensorflow 2.4.1, 11.0.221 and trained on a NVIDIA GeForce GTX 1660 GPU with batches of 225 patches. The loss function was comprised of a reconstruction part (mean squared error) and a similarity measure (cosine similarity) weighted by a coefficient α=0.005. The loss function was minimized using Adam [9].

3.3 Performance Evaluation

The percentage of abnormal voxels found in the thresholded reconstruction error maps was employed to classify them as healthy or pathological (PD) based on a threshold. The critical choice of the threshold was investigated using a Receiver Operator Curve (ROC), taking into account the imbalanced nature of our test set (15 healthy and 129 PD). Every point in the ROC corresponds to the sensitivity and specificity values obtained by a given threshold. As proposed in [15], the choice of the optimal threshold, referred to as the *pathological threshold*, was based on the optimal geometric mean, $g\text{-}mean = \sqrt{Sensitivity \times Specificity}$.

Additionally, to help evaluate the localization of anomalies, two atlases were considered: the Neuromorphometrics atlas [2] and the MNI PD25 atlas [20]. The first was used to segment the brain into 8 macro-regions: subcortical structures, white matter and the 5 gray matter lobes (Frontal, Temporal, Parietal, Occipital, Cingulate/Insular). The latter was specifically designed for PD patients exploration. It contains 8 regions: substantia nigra (SN), red nucleus (RN), subthalamic nucleus (STN), globus pallidus interna and externa (GPi, GPe), thalamus, putamen and caudate nucleus. For all of the before-mentioned regions of interest (ROI), we calculated the *g-mean* for the associated pathological threshold, leading to the classification performance of our models.

4   Results

As it can be seen in Fig. 2, both auto-encoder architectures achieve good quality reconstructions, however the SAE seems to capture finer details and textures than the AE. This explains the high contrast in AE reconstructions.

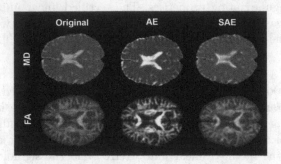

Fig. 2. Showcase of a slice of the original data and its AE and SAE reconstructions

The visualization of the percentage of abnormal voxels in the ROIs presented in Fig. 3 showcases the inter-subject variability amongst members of the same population (healthy and PD). Even so, abnormal voxels are clearly more numerous in the PD patient population.

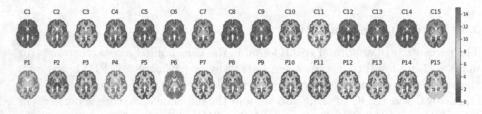

Fig. 3. The percentage of abnormal voxels found by the SAE in the anatomical ROIs presented in Sect. 3. Top: the test controls of Sample 1; Bottom: 15 randomly selected PD patients.

The *g-mean classification* scores for all models, obtained for each ROI and each sub-population, are presented in Fig. 4. We notice that, on average, the SAE model performed better than the AE on the whole brain and most of the macro and subcortical structures studied, with the exception of the temporal lobe, the putamen, the thalamus and the internal and external segment of the globus pallidus. We note that the results varied greatly across the ten populations samples. As an indication, for the whole brain, the SAE obtained a *g-mean* average score of $66.9 \pm 5.8\%$ and the AE $65.3 \pm 7.5\%$, however the best scores among the 10 samples were of 79.9% and 81.9% for the AE and SAE respectively, both on sample 1. Corresponding values for the white matter are of 68.2 ± 4.6% for SAE and 66.2±6.7% for AE. The largest standard deviations in the observed anatomical regions belonged to the white matter, the frontal and occipital lobes.

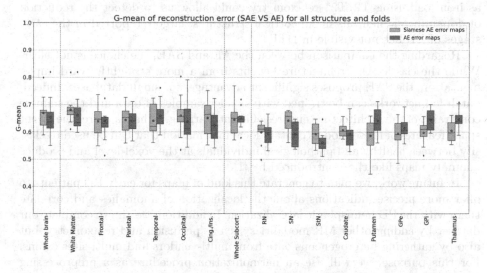

Fig. 4. *g-mean* scores for the whole brain and several ROIs for AE and SAE. The vertical dashed lines separate macro- and micro brain structures.

5 Discussion and Conclusion

Unsupervised auto-encoders (AE) have shown to efficiently tackle challenging detection tasks where brain alteration are barely seen or not visible. The objective of our study was to explore the potential of such AE models for the detection of subtle anomalies in *de novo* PD data and compare patch-based versus image-based models.

Both AE and SAE architectures produced good quality reconstructions and were able to discriminate between healthy individuals and recently diagnosed PD patients with performances (see Figs. 3 and 4) that are competitive with those found in the literature. Notably, the Correia & al. [5] SVM mean accuracy score for a selection of WM regions is of 61.3% whereas both the SAE and AE achieved g-mean scores above 66% for the WM. Also, the cross-validation procedure of Schuff & al. [18] obtained a ROC AUC of 59% for the rostral segment of the SN which is below our SN average g-mean score for the SAE and equal to that of the AE.

Note that at this early stage of the disease (1–2 H-Y scale) the patients have no tremor nor uncontrolled movements compared to healthy subjects. This rules out that movement was the index that allowed PD classification.

Using DTI data we did not search for structural atrophy or lesion load but for degradation of WM properties in the early stages of PD that could appear everywhere in the brain. This explains why the WM obtains the highest g-mean scores. This being said, our models could largely improve by increasing the size of our dataset. Furthermore, the addition of another MR modality such as iron load using T2/T2* relaxometry could allow us to detect the reduction of dopaminergic neurons in subcortical structures, largely reported in the early stages of PD but not visible in DTI.

Regarding the comparison between the AE and SAE, the choice is not clear. While the classic AE architecture benefits from a more straightforward implementation, the SAE proposes significant advantages for small databases. Indeed, patch-fed networks can be trained with smaller samples of data and the siamese constraint of the architecture ensures efficient learning. What is more, the latent space features of these models contain local information that can be used to classify between healthy and pathological individuals at the voxel-level and produce anomaly maps like those introduced by [1].

In future work, we plan to generate this kind of maps for early PD patients to offer more precise indications about the localization of anomalies and correlate them with the PD hemispheric lateralization. In addition, we aim to complete our dataset by adding other MR modalities such as perfusion and relaxometry, but also by gathering heterogeneous data from multi-vendors and multi-sites exams. For this purpose, we will use an harmonization procedure as a preprocessing step (extension of DeepHarmony [6]). Finally, our 3D implementation of the SAE model is ongoing.

References

1. Alaverdyan, Z., Jung, J., Bouet, R., Lartizien, C.: Regularized siamese neural network for unsupervised outlier detection on brain multiparametric magnetic resonance imaging: application to epilepsy lesion screening. Med. Image Anal. **60**, 101618 (2020). https://doi.org/10.1016/j.media.2019.101618
2. Bakker, R., Tiesinga, P., Kötter, R.: The scalable brain atlas: instant web-based access to public brain atlases and related content. Neuroinformatics **13**(3), 353–366 (2015). https://doi.org/10.1007/s12021-014-9258-x
3. Baur, C., Denner, S., Wiestler, B., Navab, N., Albarqouni, S.: Autoencoders for unsupervised anomaly segmentation in brain MR images: a comparative study. Med. Image Anal. **69**, 101952 (2021). https://doi.org/10.1016/j.media.2020.101952
4. Commowick, O., et al.: Objective evaluation of multiple sclerosis lesion segmentation using a data management and processing infrastructure. Sci. Rep. **8**, 13650 (2018). http://portal.fli-iam.irisa.fr/msseg-challenge
5. Correia, M.M., et al.: Towards accurate and unbiased imaging-based differentiation of Parkinson's disease, progressive supranuclear palsy and corticobasal syndrome. Brain Commun. (2020)
6. Dewey, B.E., et al.: DeepHarmony: a deep learning approach to contrast harmonization across scanner changes. Magn. Reson. Imaging (2019). https://doi.org/10.1016/j.mri.2019.05.041. Jul

7. Goodfellow, I.J., et al.: Generative adversarial networks. arXiv:1406.2661 [cs, stat], June 2014
8. Ioffe, S., Szegedy, C.: Batch normalization: accelerating deep network training by reducing internal covariate shift. arXiv:1502.03167 [cs], March 2015
9. Kingma, D.P., Ba, J.: Adam: a method for stochastic optimization. arXiv:1412.6980 [cs], January 2017
10. Kingma, D.P., Welling, M.: Auto-encoding variational bayes. arXiv:1312.6114 [cs, stat], May 2014
11. Marek, K., et al.: The parkinson's progression markers initiative (PPMI) - establishing a PD biomarker cohort. Ann. Clin. Transl. Neurol. 1460–1477 (2018)
12. MICCAI: Mild traumatic brain injury outcome prediction (2016). www.tbichallenge.wordpress.com
13. MICCAI: Ischemic stoke lesion segmentation challenge (2018). www.isles-challenge.org
14. MICCAI: Brain tumor segmentation challenge (2020). http://braintumorsegmentation.org/
15. Muñoz-Ramírez, V., Kmetzsch, V., Forbes, F., Meoni, S., Moro, E., Dojat, M.: Subtle anomaly detection in MRI brain scans: application to biomarkers extraction in patients with de novo parkinson's disease. medRxiv (2021). https://doi.org/10.1101/2021.06.03.21258269
16. Muñoz-Ramírez, V., Kmetzsch, V., Forbes, F., Dojat, M.: Deep learning models to study the early stages of parkinson's disease. In: 2020 IEEE 17th International Symposium on Biomedical Imaging (ISBI), pp. 1534–1537 (2020). https://doi.org/10.1109/ISBI45749.2020.9098529
17. Poldrack, R.A., Huckins, G., Varoquaux, G.: Establishment of best practices for evidence for prediction: a review. JAMA Psychiatry. 534–540 (2019)
18. Schuff, N., et al.: Diffusion imaging of nigral alterations in early Parkinson's disease with dopaminergic deficits. Mov. Disord. 30, 1885–1892 (2015)
19. Shinde, S., et al.: Predictive markers for Parkinson's disease using deep neural nets on neuromelanin sensitive MRI. NeuroImage: Clin. 22, 101748 (2019)
20. Xiao, Y., et al.: Multi-contrast unbiased MRI atlas of a Parkinson's disease population. Int. J. Comput. Assist. Radiol. Surg. 10, 329–341 (2015)
21. Zhao, Y.J., et al.: Progression of Parkinson's disease as evaluated by Hoehn and Yahr stage transition times. Mov. Disord. 25(6), 710–716 (2010). https://doi.org/10.1002/mds.22875

MRI Image Registration Considerably Improves CNN-Based Disease Classification

Malte Klingenberg[1,2], Didem Stark[1,2]($\boxtimes$), Fabian Eitel[1,2], and Kerstin Ritter[1,2]
for the Alzheimer's Disease Neuroimaging Initiative

[1] Department of Psychiatry and Neurosciences | CCM, Charité – Universitätsmedizin Berlin (corporate member of Freie Universität Berlin, Humboldt-Universität zu Berlin, and Berlin Institute of Health), Berlin, Germany
[2] Bernstein Center for Computational Neuroscience, Berlin, Germany

Abstract. Machine learning methods have many promising applications in medical imaging, including the diagnosis of Alzheimer's Disease (AD) based on magnetic resonance imaging (MRI) brain scans. These scans usually undergo several preprocessing steps, including image registration. However, the effect of image registration methods on the performance of the machine learning classifier is poorly understood. In this study, we train a convolutional neural network (CNN) to detect AD on a dataset preprocessed in three different ways. The scans were registered to a template either linearly or nonlinearly, or were only padded and cropped to the needed size without performing image registration. We show that both linear and nonlinear registration increase the balanced accuracy of the classifier significantly by around 6–7% in comparison to no registration. No significant difference between linear and nonlinear registration was found. The dataset split, although carefully matched for age and sex, affects the classifier performance strongly, suggesting that some subjects are easier to classify than others, possibly due to different clinical manifestations of AD and varying rates of disease progression. In conclusion, we show that for a CNN detecting AD, a prior image registration improves the classifier performance, but the choice of a linear or nonlinear registration method has only little impact on the classification accuracy and can be made based on other constraints such as computational resources or planned further analyses like the use of brain atlases.

Keywords: Alzheimer · CNN · Image registration · MRI · Deep learning

M. Klingenberg and D. Stark—These authors contributed equally to this work.
Alzheimer's Disease Neuroimaging Initiative—Data used in preparation of this article were obtained from the Alzheimer's Disease Neuroimaging Initiative (ADNI) database (adni.loni.usc.edu). As such, the investigators within the ADNI contributed to the design and implementation of ADNI and/or provided data but did not participate in analysis or writing of this report. A complete listing of ADNI investigators can be found at: http://adni.loni.usc.edu/wp-content/uploads/how_to_apply/ADNI_Acknowledgement_List.pdf.

A. Abdulkadir et al. (Eds.): MLCN 2021, LNCS 13001, pp. 44–52, 2021.
https://doi.org/10.1007/978-3-030-87586-2_5

1   Introduction

In the last years, machine learning techniques have frequently been used to diagnose neurological and psychiatric diseases, such as Alzheimer's disease (AD), based on structural magnetic resonance imaging (MRI) data [12,14,18]. While traditional methods such as support vector machines or random forests usually require a prior image registration in combination with feature extraction (e.g., volumes of brain areas or cortical thickness), convolutional neural networks (CNNs) can directly operate on 3-dimensional MRI data [9]. However, due to computational reasons and relatively small sample sizes, most studies so far focused on MRI data registered to a template (e.g., in MNI space) [18]. In addition, image registration makes it possible to compare scans taken from multiple subjects and identify relevant brain regions for a classifier [4].

Algorithms for image registration include both simple approaches that perform the registration linearly, using only translation, rotation, and scaling operations to match the input image to the reference image as well as more complicated methods that use an additional nonlinear warping step to achieve a better correspondence between the two images. The latter increases the computational cost, but also improves the image similarity. While there are studies comparing different image registration methods based on similarity metrics such as the overlap between specific volumes in the input and reference images, the similarity between these volumes, or differences in their boundaries [1,2,7,13,16], to the best of our knowledge, there has not been an analysis on the effect of the registration method on the performance of a machine learning classifier.

In the present study, we examine how a CNN performs in AD classification when trained on a dataset preprocessed in three different ways, namely no registration but only image cropping, linear registration, and nonlinear registration. The CNN receives the entire 3D scan as input, without prior extraction of regions of interest or other pre-selection of image features. To eliminate potential confounders, we carefully balanced the data set with respect to age and sex, a step often neglected in previous work.

2   Data and Methods

2.1   Dataset

Data used in the preparation of this article were obtained from the Alzheimer's Disease Neuroimaging Initiative (ADNI) database (adni.loni.usc.edu). The ADNI was launched in 2003 as a public-private partnership, led by Principal Investigator Michael W. Weiner, MD. The primary goal of ADNI has been to test whether serial magnetic resonance imaging (MRI), positron emission tomography (PET), other biological markers, and clinical and neuropsychological assessment can be combined to measure the progression of mild cognitive impairment (MCI) and early AD.

We included subjects from all ADNI study phases who were, at their baseline visit, either classified as cognitively normal (CN) or diagnosed with AD. Subjects who were classified as CN at their baseline visit, but at some later visit received a diagnosis of MCI or AD were excluded from our analysis. Overall, our study population included 573 subjects (406 CN, 167 AD). Since the mean age of female subjects was less than the mean age of the male subjects in both groups (AD: 74.0 ± 7.9 to 75.2 ± 8.3 years, $p = 0.3211$; CN: 71.3 ± 6.7 to 73.6 ± 6.8 years, $p = 0.0007$; p-values were calculated with a two-sample t-test), we removed this possible source of bias by undersampling based on subject sex. We divided the population into bins according to their diagnosis (AD or CN) and their age (in 5-year ranges, e.g. 60–64, 65–69 etc.) and if such a bin contained a different number of female and male subjects, we randomly removed subjects from the bin until their number was equal. In total, this process reduced the population size by about 25%, with the final population consisting of 432 subjects (306 CN, 126 AD). The resulting female and male age distributions were similar, with their means not significantly different anymore (p > 0.75).

The population was then split into a training set (306 subjects) and validation and test sets (63 subjects each). We used a stratified split based on subject sex, diagnosis, and age range to ensure that the distributions of the sets are as close as possible. It is important to split the dataset on the subject level instead of the image level to avoid data leakage [18], as otherwise scans of a single subject may end up in both the training and the test set.

The final size of the training set ranged from 758 to 834 images, depending on the number of scans taken from the specific subjects remaining after the undersampling. For the test set, we only kept the scans taken at the baseline visit. As the resulting test set was rather small, the results were likely to vary significantly depending on the specific data split. To avoid the possible issue of a lucky or unlucky split, we repeated both the undersampling and the data split for ten different random seeds. The subsequent analyses were then performed on all ten resulting data splits.

2.2 Image Preprocessing

For this analysis, we used T1-weighted structural MRI scans acquired at a magnetic field strength of 3 T. The scans were taken at different imaging sites and had undergone gradwarping (gradient inhomogeneity correction), intensity correction, and were scaled for gradient drift using the phantom data.

We preprocessed the scans in three different ways, using the 1mm T1 version of the ICBM152 reference brain as a template. First, if the slice thickness of a scan in any plane was different from 1mm, the scan was resampled to that resolution. We then (1) used Advanced Normalization Tools (ANTs)[1] to linearly register the scans to the template (transform type parameter 'a'); (2) additionally used the ANTs implementation of the SyN algorithm [3] to apply nonlinear warping to better fit the template (transform type parameter 's'); and (3) as a baseline comparison, only padded and/or cropped the scan to fit the dimensions of the template without performing any image registration. We have chosen the

---

[1] http://stnava.github.io/ANTs/.

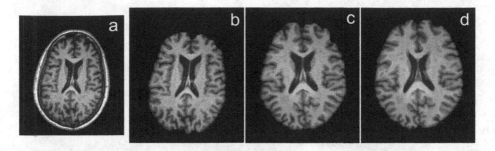

Fig. 1. Axial slice of a sample raw scan (a) and the resulting scans after preprocessing in three different ways: padding/cropping to the template dimensions (b), linear registration (c) and nonlinear registration (d). In all three cases, skull-stripping followed. Note the differing dimensions of the raw scan ($208 \times 240 \times 256$) and the preprocessed scans ($182 \times 218 \times 182$). The slightly off-center positioning and rotation of the brain in the raw scan is retained in the padded/cropped scan (b), while the brain is centered in the template and thus in the two registered scans (c) and (d).

SyN algorithm because of its superior performance reported in [13]. In all cases, we then used the FSL Brain Extraction Tool (fsl-bet) for skull stripping [11,17]. Figure 1 shows a comparison of the preprocessing results.

2.3 Network Architecture and Training

The CNN architecture chosen for this paper was taken from [4] and contains four convolutional blocks, each consisting of a convolutional layer with filter size 3 and 8, 16, 32, and 64 features respectively, as well as batch normalization and max pooling with window size 2, 3, 2 and 3. The convolutional blocks are followed by a fully connected layer of 128 units and the 2-unit output layer (representing the two classes CN and AD). Before each of these two layers, dropout is applied with a value of $p = 0.4$.

Training the network was done using the Adam optimizer and cross entropy loss, with the learning rate and weight decay set to 10^{-4}. The chosen batch size of 16 was limited by the available GPU memory. The training data was augmented by translating the scans along the sagittal axis by up to 2 voxels and mirroring the scans across the sagittal plane. When the balanced accuracy achieved by the model on the validation set did not improve for 8 epochs, training was stopped. The model with the best balanced accuracy on the validation set was then evaluated on the test set. The training process was repeated five times for each split and the results averaged for each of the ten splits, to increase the robustness of the results.

3 Results

The performance achieved by the classifier on the three differently processed datasets is shown in Fig. 2. The best results are obtained using the nonlinear

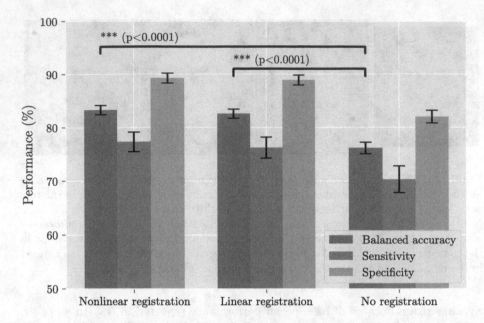

Fig. 2. Mean balanced accuracy, sensitivity, and specificity achieved by the classifier on the test set across all runs for all splits. The error bars show the standard error. Significance values were calculated using the Wilcoxon signed-rank test.

preprocessing, with a balanced test accuracy mean and standard error of $83.3 \pm 0.9\%$ (training accuracy $93.1 \pm 0.5\%$, validation accuracy $87.2 \pm 0.7\%$). The linear preprocessing performs only slightly worse at $82.6 \pm 0.9\%$ (training $94.4 \pm 0.3\%$, validation $87.8 \pm 0.5\%$), while the classifier trained on unregistered data achieves a balanced accuracy of $76.2 \pm 1.1\%$ (training $85.3 \pm 0.6\%$, validation $79.4 \pm 0.8\%$). The improvement in balanced test accuracy for either registration method over the unregistered data is significant ($p < 0.0001$, Wilcoxon signed-rank test), but the difference between linear and nonlinear registration is not ($p = 0.2885$).

We also calculated the receiver operating characteristic (ROC) curve for each individual classifier and then averaged them for each preprocessing method. The average ROC curves are shown in Fig. 3. Again, the results for nonlinear and linear registration are very similar, with an area under the curve (AUC) of 0.910 ± 0.008 and 0.904 ± 0.008 respectively, while the classifier using non-registered images performs slightly worse at 0.850 ± 0.011. Additionally, there was a very strong dependence of the classifier performance on the specific dataset split the model was trained on. Figure 4 shows the mean balanced accuracy for each split, with a difference between splits of as much as 17%. The observed effect still holds, with the classifier trained on unregistered images achieving the worst results in all but one split. For most splits, using nonlinear registration gives the best results, while linear registration outperforms nonlinear registration in two of the ten splits.

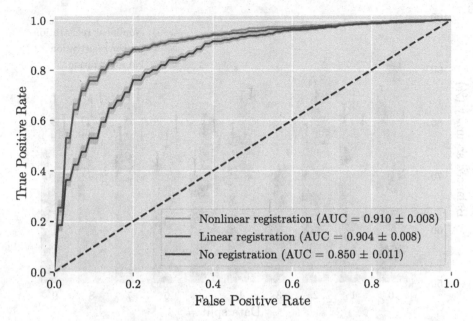

Fig. 3. Average receiver operating characteristic (ROC) curves for the three prepro-cessing methods. The shaded area shows the standard error of the curve.

4   Discussion

In this study, we have trained a 3D CNN on MRI brain scans for an AD vs. CN classification task using a dataset balanced for sex and age. We preprocessed the dataset in three different ways: no registration but only image cropping, linear registration, and nonlinear registration. Registering the input images improves the classifier performance by 6–7%, depending on the registration method. While image registration is common practice, this result is not obvious, as the infor-mation removed by registering the images to a template could be relevant for the classification task itself. Linear registration already eliminates all but slight differences between the inputs in positioning, rotation, and size of the brain within the scan. Nonlinear registration goes further by not only matching the brain as a whole to the template but also matching individual structures result-ing in the removal of relative sizes of specific brain areas [1]. Both registration methods remove variations and result in a more homogenized input [7,13]. More recently, deep learning based image registration methods are also proposed, how-ever, those methods have not yet become part of publicly available medical image processing tools [10,16].

Image registration can impact classification performance both negatively or positively, depending on how relevant the variations in the input were. In our case, registering the images benefits the classification. The information that was removed therefore seems to not have been salient, instead masking the actually important image features. The structural changes in the brain brought about

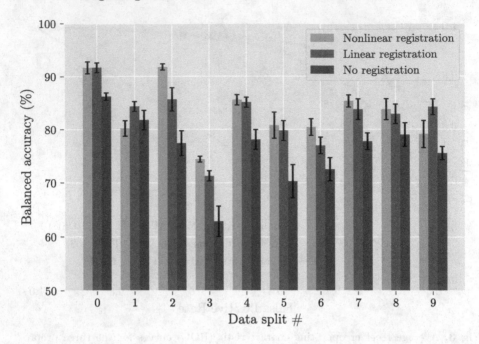

Fig. 4. Mean balanced accuracy achieved by the classifier on the different splits. The error bars show the standard error.

by AD are subtle, especially in milder cases, and on a smaller scale than the variations in brain positioning and size present in the input images. Removing these large-scale variations therefore enables the network to better focus on the smaller variations important for deciding the classification task.

While there is a benefit to using registered instead of non-registered images, we did not find a significant difference between linear and nonlinear registration. This suggests that, while large-scale variations like subject positioning impede the performance of the network, it is not dependent on the more precise alignment to the template achieved by using a nonlinear registration method. Because the deformations introduced during nonlinear registration are usually small, this might be explained by the invariance to small shifts and distortions caused by the pooling layers of the CNN [15]. For a thorough region-wise analysis across subjects as for example done in [4,8], however, a nonlinear registration still can be beneficial.

We would like to point out the following limitations and ideas for future work. First, although we compared for the first time the effect of different registration methods on a machine learning classifier, future studies should investigate whether our results generalize to different software packages performing linear and nonlinear registration, and different tasks and data sets. Moreover, it should be investigated whether the difference between no registration and linear or nonlinear registration can be alleviated in a larger data set ($>$10.000 subjects), allowing to learn across large-scale variations. Second, although we achieved a balanced accuracy on par with results reported in the literature [18], the performance of our classifier could likely have been improved with a thorough hyperparameter optimization. While the training, validation and test accuracies are reasonably close, further increases might be possible. However, since we have run extensive repetitions in order to get robust results, hyperparameter optimization was not possible due to computational time constraints. And third, although the conclusions about the superiority of registration over no registration remain, we have found a strong dependence of the results on the specific dataset split used in training the classifier, suggesting that some subjects are generally easier to correctly classify than others. This is supported by the fact that there are splits for which all preprocessing methods give good results, while for other splits all methods perform rather poorly, see for example splits 0 and 3 in Fig. 4. These differences are therefore likely caused by the data, i.e. the subjects themselves rather than by other effects or random chance. As the varying performance for different splits may also in part be caused by the small test set size, future research should address this problem by, for example, using oversampling instead of undersampling to balance the dataset. While this can lead to overfitting, using oversampling instead of undersampling or combining the two approaches would increase the test set size and has been shown to be capable of improving classifier performance [5,6].

Acknowledgements. We thank Tobias Scheffer for his useful suggestions. We acknowledge support from the German Research Foundation (DFG, 389563835; TRR 265, 402170461; CRC 1404, 414984028), the Brain & Behavior Research Foundation (NARSAD Young Investigator Grant, USA) and the Manfred and Ursula-Müller Stiftung.

Data collection and sharing for this project was funded by the Alzheimer's Disease Neuroimaging Initiative (ADNI) (National Institutes of Health Grant U01 AG024904) and DOD ADNI (Department of Defense award number W81XWH-12-2 0012). ADNI is funded by the National Institute on Aging, the National Institute of Biomedical Imaging and Bioengineering, and through generous contributions from the following: AbbVie, Alzheimer's Association; Alzheimer's Drug Discovery Foundation; Araclon Biotech; BioClinica, Inc.; Biogen; Bristol-Myers Squibb Company; CereSpir, Inc.; Cogstate; Eisai Inc.; Elan Pharmaceuticals, Inc.; Eli Lilly and Company; EuroImmun; F. Hoffmann-La Roche Ltd and its affiliated company Genentech, Inc.; Fujirebio; GE Healthcare; IXICO Ltd.; Janssen Alzheimer Immunotherapy Research & Development, LLC.; Johnson & Johnson Pharmaceutical Research & Development LLC.; Lumosity; Lundbeck; Merck & Co., Inc.; Meso Scale Diagnostics, LLC.; NeuroRx Research; Neurotrack Technologies; Novartis Pharmaceuticals Corporation; Pfizer Inc.; Piramal Imaging; Servier; Takeda Pharmaceutical Company; and Transition Therapeutics. The Canadian Institutes of Health Research is providing funds to support ADNI clinical sites in Canada. Private sector contributions are facilitated by the Foundation for the National Institutes of Health (www.fnih.org). The grantee organization is the Northern California Institute for Research and Education, and the study is coordinated by the Alzheimer's Therapeutic Research Institute at the University of Southern California. ADNI data are disseminated by the Laboratory for Neuro Imaging at the University of Southern California.

References

1. Abderrahim, M., Baâzaoui, A., Barhoumi, W.: Comparative study of relevant methods for MRI/X brain image registration. In: Jmaiel, M., Mokhtari, M., Abdulrazak, B., Aloulou, H., Kallel, S. (eds.) ICOST 2020. LNCS, vol. 12157, pp. 338–347. Springer, Cham (2020). https://doi.org/10.1007/978-3-030-51517-1_30
2. Andrade, N., Faria, F.A., Cappabianco, F.A.M.: A practical review on medical image registration: From rigid to deep learning based approaches. In: 2018 31st SIBGRAPI Conference on Graphics, Patterns and Images, pp. 463–470 (2018)
3. Avants, B.B., Epstein, C.L., Grossman, M., Gee, J.C.: Symmetric diffeomorphic image registration with cross-correlation: evaluating automated labeling of elderly and neurodegenerative brain. Med. Image Anal. 12(1), 26–41 (2008)
4. Böhle, M., Eitel, F., Weygandt, M., Ritter, K.: Layer-wise relevance propagation for explaining deep neural network decisions in MRI-based Alzheimer's disease classification. Front. Aging Neurosci. 11, 194 (2019)
5. Buda, M., Maki, A., Mazurowski, M.A.: A systematic study of the class imbalance problem in convolutional neural networks. Neural Netw. 106, 249–259 (2018)
6. Chawla, N.V., Bowyer, K.W., Hall, L.O., Kegelmeyer, W.P.: Smote: synthetic minority over-sampling technique. J. Artif. Intell. Res. 16, 321–357 (2002)
7. Dadar, M., Fonov, V.S., Collins, D.L., Alzheimer's Disease Neuroimaging Initiative: A comparison of publicly available linear MRI stereotaxic registration techniques. NeuroImage 174, 191–200 (2018)
8. Eitel, F., Ritter, K., for the Alzheimer's Disease Neuroimaging Initiative (ADNI): Testing the robustness of attribution methods for convolutional neural networks in MRI-based Alzheimer's disease classification. In: Suzuki, K., et al. (eds.) MLCDS/IMIMIC -2019. LNCS, vol. 11797, pp. 3–11. Springer, Cham (2019). https://doi.org/10.1007/978-3-030-33850-3_1
9. Eitel, F., Schulz, M.A., Seiler, M., Walter, H., Ritter, K.: Promises and pitfalls of deep neural networks in neuroimaging-based psychiatric research. Exp. Neurol. 339, 113608 (2021)
10. Haskins, G., Kruger, U., Yan, P.: Deep learning in medical image registration: a survey. Mach. Vis. Appl. 31(1), 1–18 (2020)
11. Jenkinson, M., Beckmann, C.F., Behrens, T.E., Woolrich, M.W., Smith, S.M.: FSL. Neuroimage 62(2), 782–790 (2012)
12. Jo, T., Nho, K., Saykin, A.J.: Deep learning in Alzheimer's disease: diagnostic classification and prognostic prediction using neuroimaging data. Front. Aging Neurosc. 11, 220 (2019)
13. Klein, A., et al.: Evaluation of 14 nonlinear deformation algorithms applied to human brain MRI registration. NeuroImage 46(3), 786–802 (2009)
14. Klöppel, S., et al.: Accuracy of dementia diagnosis–a direct comparison between radiologists and a computerized method. Brain 131(11), 2969–2974 (2008)
15. LeCun, Y., Bengio, Y., Hinton, G.: Deep learning. Nature 521(7553), 436–444 (2015)
16. Nazib, A., Fookes, C., Perrin, D.: A comparative analysis of registration tools: traditional vs deep learning approach on high resolution tissue cleared data. arXiv preprint arXiv:1810.08315 (2018)
17. Smith, S.M.: Fast robust automated brain extraction. Hum. Brain Mapp. 17(3), 143–155 (2002)
18. Wen, J., et al.: Convolutional neural networks for classification of Alzheimer's disease: overview and reproducible evaluation. Med. Image Anal. 63, 101694 (2020)

Dynamic Sub-graph Learning for Patch-Based Cortical Folding Classification

Zhiwei Deng, Jiong Zhang, Yonggang Shi$^{(\boxtimes)}$,
and the Health and Aging Brain Study (HABS-HD) Study Team

USC Stevens Neuroimaging and Informatics Institute, Keck School
of Medicine of USC, University of Southern California, Los Angeles, USA
yshi@loni.usc.edu

Abstract. Surface mapping techniques have been commonly used for the alignment of cortical anatomy and the detection of gray matter thickness changes in Alzheimer's disease (AD) imaging research. Two major hurdles exist in further advancing the accuracy in cortical analysis. First, high variability in the topological arrangement of gyral folding patterns makes it very likely that sucal area in one brain will be mapped to the gyral area of another brain. Second, the considerable differences in the thickness distribution of the sulcal and gyral area will greatly reduce the power in atrophy detection if misaligned. To overcome these challenges, it will be desirable to identify brains with cortical regions sharing similar folding patterns and perform anatomically more meaningful atrophy detection. To this end, we propose a patch-based classification method of folding patterns by developing a novel graph convolutional neural network (GCN). We focus on the classification of the precuneus region in this work because it is one of the early cortical regions affected by AD and considered to have three major folding patterns. Compared to previous GCN-based methods, the main novelty of our model is the dynamic learning of sub-graphs for each vertex of a surface patch based on distances in the feature space. Our proposed network dynamically updates the vertex feature representation without overly smoothing the local folding structures. In our experiments, we use a large-scale dataset with 980 precuneus patches and demonstrate that our method outperforms five other neural network models in classifying precuneus folding patterns.

Keywords: Cortical folding analysis · Alzheimer disease · Graph convolutional neural network

HABS-HD MPIs: Sid E O'Bryant, Kristine Yaffe, Arthur Toga, Robert Rissman, & Leigh Johnson; and the HABS-HD Investigators: Meredith Braskie, Kevin King, James R Hall, Melissa Petersen, Raymond Palmer, Robert Barber, Yonggang Shi, Fan Zhang, Rajesh Nandy, Roderick McColl, David Mason, Bradley Christian, Nicole Philips and Stephanie Large.

© Springer Nature Switzerland AG 2021
A. Abdulkadir et al. (Eds.): MLCN 2021, LNCS 13001, pp. 53–62, 2021.
https://doi.org/10.1007/978-3-030-87586-2_6

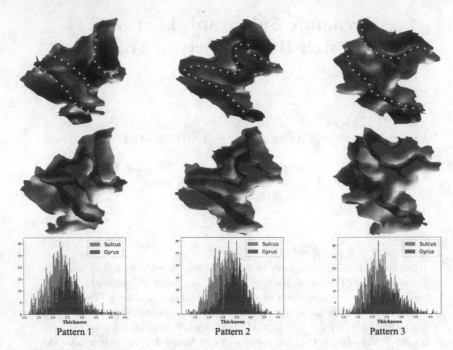

Fig. 1. An illustration of three major folding patterns of the precunues cortical region. Pattern 1: three approximately parallel gyri; Pattern 2: M-shape gyri; Pattern 3: two approximately parallel gyri. Row 1: precuneus patches colored by the principal curvature; Row 2: segmentation of the patch into gyral (red) and sulcal (yellow) areas. Row 3: gray matter thickness distribution of the gyral and sulcal area of each patch. (Color figure online)

1 Introduction

To map cortical atrophy in Alzheimer's disease (AD), conventional approaches relied on surface mapping techniques that computed the assumed one-to-one correspondences across different brains [6,19]. The high variability of the cortical folding patterns across subjects, however, have been well studied and known in brain anatomy [3,16]. In addition, there are salient differences in gray matter thickness between the sulci and gyri of the cortical ribbon [5]. Taken together, these two factors make it difficult for mapping-based methods to establish meaningful correspondences in many association cortices critical for AD diagnosis because they could likely mix sulcal and gyral areas with different thickness distributions. To overcome this fundamental limitation, we propose to develop a novel graph convolutional network (GCN) for patch-based cortical folding classification. Our goal is to enable the mapping of cortical patches with similar folding patterns and ultimately enhance the power in detecting localized cortical atrophy in brain degeneration.

While the variability of gyral and sulcal folding patterns over the whole cortex could be immense across the population, the number of folding patterns in each cortical region can be more tractable. In this work, we will thus focus on a patch-based approach to develop our cortical folding classification method. As one of the early cortical regions affected by AD, the precuneus region exhibits three main folding patterns [16] and is thus a great test bed for our method development and evaluation. As shown in Fig. 1 there are three different folding patterns of the precuneus region with different gray matter thickness distributions between the sulcal and gyral areas. With these heterogeneous topological patterns of the gyri and sulci, surface mapping methods cannot avoid to match some of the gyral area with the sulcal portion of the precuneus across different subjects if different cortical folding patterns are not classified and mapped separately.

Deep learning has been applied on many neuroimaging tasks and achieved human-level performances for structure classification [1,9,13,17,18]. To learn the geometrical features, graph-based convolution networks (GCN) have been proposed and proven effective in shape analysis tasks [1,9], where graph convolution operator [11] was designed to simulate the convolution operator in CNNs. Unlike conventional convolutional operators on fixed grids, graph convolution operators learn the graph representations through feature propagation along connected edges, which enables the frameworks to learn from local to global features. Typically, the propagation graph is fixed once it is constructed, which limits the propagation of vertices that are close in feature spaces but far in the geometric space. In addition, the global propagation process of GCN could also smooth node features across sulcal and gyral areas in our problem and obscure discriminative features.

To alleviate this problem, in this paper, we propose a novel dynamic sub-graph propagation model for the classification of the folding pattern in precuneus cortical patches. As illustrated in Fig. 2, the proposed method has 2 critical components: (1) a dynamic sub-graph construction strategy, which can aggregate vertices with similar features to prevent the over-smoothing problem during the propagation process. (2) a graph-based vertex representation learning module via propagation within the sub-graphs. During the learning process, these components allow each layer of the network to learn the optimal graph structure to generate the vertex feature representation for the final classification. In our experiments, we applied our method to a large-scale dataset of 980 precuneus patches and demonstrate that our novel model achieved superior classification accuracy than five other neural network methods.

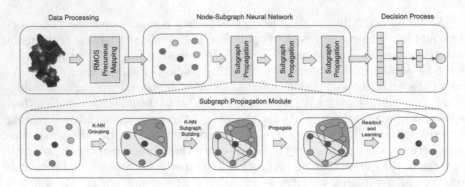

Fig. 2. The architecture of the proposed dynamic sub-graph learning model.
Top row: the pipeline of the whole end-to-end model, which pre-processes and learns
different structural pattern information by minimizing the Cross-entropy loss. Bottom
row: a zoom-in view of the sub-graph propagation module, which adaptively updates
the sub-graph transition matrix of each vertex and learns the representations.

2   Methods

In this section, we develop our novel approach to classify the folding pattern of
the precuneus patches inspired by the graph convolutional neural network (GCN)
[11]. Instead of propagating node features globally by GCN, we exploit the local
features of the surface by constructing sub-graphs on each vertex of the surface
patch. Before we develop our classification method, we leverage existing surface
mapping method [7] to establish a common vertex indices for surface patches
from different subjects. This will help factor out macroscopic differences across
the cortices and allow us to focus on the classification of more detailed folding
patterns. Given a precuneus patch $P(V)$, where $V = \{v_1, v_2, ..., v_n\}$ denotes a set
of vertices on the precuneus surface, where the vertex indices are common for all
subjects. Let $A \in \mathbb{R}^{n \times n}$ represents the connection matrix, and $X \in \mathbb{R}^{n \times d}$ be the
feature matrix, where each row of X represents the feature of the corresponding
node. Our aim is to learn a patch representation from the input node features.

Vertex Representation Learning. Consider for each vertex v_i in V, there is a
corresponding feature vector $x_i \in \mathbb{R}^d$, where d represents the feature dimension.
To better capture the geometry features of the input patch, the initial feature
dimension is set to 4 for the 1st sub-graph propagation layer as shown in Fig. 2,
which includes the 3-D coordinates and the 1-D principal curvature value at
each vertex. The d feature dimension can also be extended by adding other
shape descriptors and functional characterizations.

The node feature propagation process has been proven to be efficient for
feature learning with graph-structured data [8,11]. In general, the propagation
process can be regarded as the multiplication of the normalized connection transition matrix A and the feature matrix X^l of the l-th layer as:

$$X^{l+1} = A^T X^l \tag{1}$$

where each element a_{ij} of A indicates the connection between the vertex i and vertex j. This process can learn the updated node representation by taking both the current node feature and the features from connected nodes into consideration. However, this method propagates vertex features globally, which may overly smooth the features across the sulcal and gyral areas and hence obscure the local folding structures. Such loss of sucal vs gyral contrast may compromise the final structure learning for pattern classification.

To better learn the node features, we propose to represent local structures in sub-graph propagation layers shown in Fig. 2. In our work, instead of aggregating the features from vertices in the geometric neighborhood, we construct a sub-graph among these neighboring nodes in the feature space and use this sub-graph as the feature of the current node to characterize its local folding information. More specifically, as shown in the bottom row of Fig. 2, we first apply the K-nearest-neighbor(K-NN) algorithm to find $K1$ ($K1 = 4$ in the example in Fig. 2) neighbors $V_{N_{blue}}$ in the feature space for the blue node. This collection of neighboring vertices are shown as the yellow patch in the second row of Fig. 2. After that, we apply another K-NN search process to identify $K2$ ($K2 = 2$ in the example in Fig. 2) neighbors for every node in $V_{N_{blue}}$ to construct the sub-graph, which is the graph structure shown in the yellow patch in Fig. 2. During the propagation process on the sub-graph, features of each node in the sub-graph are updated and finally the updated sub-graph is readout and learned as input to the next sub-graph propagation layer. There are overall three sub-graph propagation layers in our model. Finally, the output from the sub-graph layers are sent to fully-connected layers to learn the cortical folding classification. The feature learning process in the sub-graph propagation layer can be expressed as:

$$X_{N_i}^{l+1} = G_i^{l+1} X_{N_i}^l \tag{2}$$

$$X_i^{l+1} = f(W \times (max(X_{N_i}^{l+1})||avg(X_{N_i}^{l+1})||sum(X_{N_i}^{l+1})) + b) \tag{3}$$

where $X_{N_i}^l$ and G_i^l represents the features of V_{N_i} and sub-graph of the i-th node that constructed using V_{N_i} at the l-th layer, max, avg, sum denote the pooling operators, and $||$ represents the concatenation operator. W is a trainable projection matrix, b is the bias term, and f denotes the activation function. X_i^{l+1} is the readout to the next layer. In every sub-graph propagation module, the same process is applied to every vertex of the input graph in parallel, which means the sub-graphs are generated only using the input features of each layer. It's worth noting that the propagation with our proposed method not only maintains the vertex features, but also includes the structural information due to the connection information is remained in the sub-graphs.

Dynamic Graph Construction. In graph convolutional neural networks [11], the node features are propagated along fixed graph edges, which is usually constructed with geometric neighborhoods. However, the nodes with similar features are not necessarily close in the geometric space, thus the information sharing between these nodes are limited under the context of GCN. In contrast from the graph convolutional neural network where the transition matrix is fixed, the sub-graphs in this work is dynamically constructed layer by layer. In our model, the union of the sub-graphs can be denoted by $G^l = \{G_1^l, G_2^l, ..., G_n^l\}$ at the l-th layer, where G_i^l is the sub-graph in V_{N_i}, the set of neighboring vertices of the i-th node constructed by the K-NN algorithm in the feature spaces of each layer. In addition, this dynamically graph construction process can potentially increase the number of *intra-class* connections and decrease the number of *inter-class* connections, which has been proven to be an efficient strategy to prevent the *over-smoothing* problem for GCN [2]. Furthermore, in graph-based shape recognition tasks, dynamically updating the propagation graph is used to enlarge the receptive field and group the points in semantic spaces (sulcal or gyral area in our problem), which enables the model to learn not only the node features, but also how to construct the graphs for better representation [20].

Table 1. Mean classification accuracy with 5-Fold cross validation

Methods	ADNI (%)	HABS-HD (%)	Combined (%)
Graph-CNN	82.1	82.2	85.0
Graph-Unet	84.6	83.3	86.3
DGCNN	85.6	85.0	86.8
DenseNet	80.4	81.6	83.5
VoxNet	71.5	67.5	70.3
Ours	**86.3**	**86.0**	**88.5**

3   Experimental Results

Datasets and Labeling. In our experiment, we use the T1-weighted MRI of 588 subjects from the Alzheimer's Disease Neuroimaging Initiative (ADNI) [14] and 1675 subjects from the Health and Aging Brain Study: Health Disparities (HABS-HD) study [15]. Cortical surfaces were first reconstructed by FreeSurfer and the precuneus patches were then extracted based on the ROI label from FreeSurfer. Following previous anatomical description [4], we manually inspected the cortical foldings of the left hemisphere of all subjects. During this screening process, we identified potentially more general folding patterns than the three-class definition proposed in [16], which could be interesting for future anatomical research. In the current experiment, however, we follow the existing anatomy literature [4] and use 980 subjects from these two cohorts that match the three-class definition. More specifically, there are 284 and 696 patches selected from the ADNI3 and the HABS-HD dataset, respectively. In the combined dataset, there are 417, 301 and 262 patches labeled as pattern 1, 2 and 3, respectively. Example patches of different patterns are shown in Fig. 1.

Experiments Setting. We perform three classification tasks to evaluate our model's overall performance on the ADNI3 (n = 284), HABS-HD (n = 696),

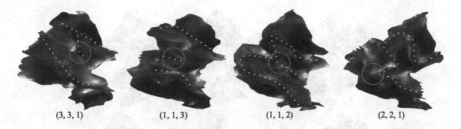

(3, 3, 1)   (1, 1, 3)   (1, 1, 2)   (2, 2, 1)

Fig. 3. A comparison of prediction performance on four difficult cases. Under each case, the (true label, our prediction, Graph-CNN prediction) are listed. The disturbances in the folding patterns are unexpected gyral branch, short gyrus, broken gyrus, and unexpected gyral extension, respectively.

and the combination of these two datasets (n = 980). For each task, we use a 5-fold cross-validation to evaluate the robustness of the models. For comparison, we selected three graph-based (Graph-CNN [18], Graph U-nets [8], and DGCNN [20]), and two voxel-based (VoxNet [12] and DenseNet [10]) neural network models. In voxel-based models, the surface patch surfaces were voxelized into a $52 \times 52 \times 52$ cube and the structural representations are learned by 3-D convolution operators. For fair comparison, the decision MLP and the convolutional layers of different models are set as the same as the proposed model. More specifically, the number of graph convolutional layers in Graph-CNN and Graph U-nets, or the EdgeConv layers in DGCNN are set to 3. The same number of 3-D convolutional layers and dense layers were applied in the VoxNet and DenseNet. For all models, ELU is used as the activation function for avoiding the dead neuron problem, and the dropout (a rate of 0.3) was applied in both fully-connected layers and GCN layers for better generalization. All experiments are implemented and tested with Pytorch and NVIDIA GTX 1080 GPU with 8 GB memory.

Results on Precuneus Classification. Experimental results of all methods are listed in Table 1. The classification accuracy obtained by our model in all three classification tasks are 86.3% , 86.0% and 88.5% respectively, which are the highest among all methods. From our experiments, we also have three important observations. First, all graph-based models generally perform better than voxel-based models. This shows that describing the cortical surface patches as graph-structured data is more efficient and robust than voxel-structured data. Second, it is worth noting that DGCNN and the proposed method are the only methods here that dynamically update the graph structure during the learning process, and both of them surpass other graph-based methods. This validates that the dynamically constructed graphs can enable the model to learn how to propagate information more effectively. Third, the proposed model has demonstrated more robust performance to atypical patterns in our experience. As shown in Fig. 3, our model is able to better handle cases with atypical patterns deviating from the three-class definition.

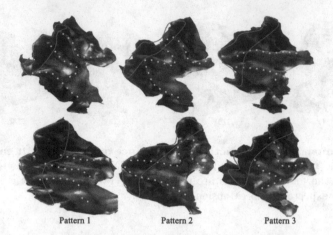

Pattern 1   Pattern 2   Pattern 3

Fig. 4. Second-order energy distribution heat map of vertex representation.
Brighter color means higher energy. Most representative energy is distributed in gyral
areas. Generally shared gyral areas enclosed by the red curves are suppressed and have
little representative energy. (Color figure online)

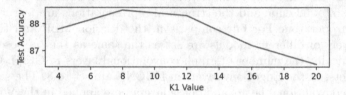

Fig. 5. Effects of neighborhood size on classification accuracy on the combined dataset
of ADNI and HABS-HD.

Analysis of Structure Learning. We also investigated the information that
the model learned to verify that our model classifies the precuneus patches based
on correct anatomical details. After the three sub-graph propagation layers in
Fig. 2, the vertex representation's energy distribution map can be computed
according to [21] and plotted for different folding patterns in Fig. 4. As can be
seen, most energy is focused on the gyral area of the patch, which is exactly
the anatomical classification criteria used in [4]. Furthermore, our model can
suppress the irrelevant gyral information in the upper area of each case (enclosed
in the red curves) to enhance the discriminative power in classification.

Effects of Neighborhood Sizes. To examine the effects of the neighborhood
construction process, we trained our model with different neighborhood sizes.
The model's classification performance is evaluated with K1 equal to 4, 8, 12,
16, 20 respectively, and K2 set to be K1/2 for sub-graph construction. The
classification accuracy are plotted in Fig. 5. The model performs best with K1 = 8
and the performance will drop if K1 is too small due to insufficient neighborhood
representation. On the other hand, the performance of the model will also decline

if K1 is too large, which might be due to increased risk of the over-smoothing problem for graph propagation as mentioned in [2]. Thus, our model is trained with the best parameter setting ($K1 = 8$ and $K2 = 4$).

4 Conclusions

In this paper, we proposed a node sub-graph network for the classification of cortical patches, which can dynamically change the graph connection of vertices and effectively learn the local folding structure information. In the proposed network, the vertex features are propagated based on automatically updated sub-graphs in each layer. In comparisons to five existing neural networks, we demonstrated the superior performance of the proposed network on a large-scale dataset of precuneus patches. For future work, we will extend the current framework to the classification of other cortical regions, and enable the ROI-based clustering of subjects in population studies for enhanced power of brain atrophy detection in AD.

Acknowledgements. This work was supported by the National Institute of Health (NIH) under grants RF1AG064584, RF1AG056573, R01EB022744, R21AG064776, P41EB015922, P30AG066530. Research reported on this publication was also supported by the National Institute on Aging of the NIH under Award Numbers R01AG054073 and R01AG058533. The content is solely the responsibility of the authors and does not necessarily represent the official views of the NIH.

References

1. Besson, P., Parrish, T., Katsaggelos, A.K., Bandt, S.K.: Geometric deep learning on brain shape predicts sex and age. bioRxiv (2020)
2. Chen, D., Lin, Y., Li, W., Li, P., Zhou, J., Sun, X.: Measuring and relieving the over-smoothing problem for graph neural networks from the topological view (2019)
3. Ding, S.L., Van Hoesen, G.W.: Borders, extent, and topography of human perirhinal cortex as revealed using multiple modern neuroanatomical and pathological markers. Hum. Brain Mapp. **31**(9), 1359–1379 (2010)
4. Duan, D., et al.: Exploring folding patterns of infant cerebral cortex based on multi-view curvature features: methods and applications. NeuroImage **185**, 575–592 (2019)
5. Fischl, B., Dale, A.M.: Measuring the thickness of the human cerebral cortex from magnetic resonance images. Proc. Natl. Acad. Sci. **97**(20), 11050–11055 (2000)
6. Fischl, B., Sereno, M.I., Dale, A.M.: Cortical surface-based analysis: Ii: Inflation, flattening, and a surface-based coordinate system. NeuroImage **9**(2), 195–207 (1999)
7. Gahm, J.K., Tang, Y., Shi, Y.: Patch-based mapping of transentorhinal cortex with a distributed atlas. In: Frangi, A.F., Schnabel, J.A., Davatzikos, C., Alberola-López, C., Fichtinger, G. (eds.) MICCAI 2018. LNCS, vol. 11072, pp. 689–697. Springer, Cham (2018). https://doi.org/10.1007/978-3-030-00931-1_79
8. Gao, H., Ji, S.: Graph u-nets. In: Chaudhuri, K., Salakhutdinov, R. (eds.) Proceedings of the 36th International Conference on Machine Learning. Proceedings of Machine Learning Research, vol. 97, pp. 2083–2092. PMLR 09–15 June 2019

9. Gopinath, K., Desrosiers, C., Lombaert, H.: Graph convolutions on spectral embeddings for cortical surface parcellation. Med. Image Anal. **54**, 297–305 (2019)
10. Gottapu, R.D., Dagli, C.H.: Densenet for anatomical brain segmentation. Procedia Comput. Sci. **140**, 179–185 (2018). cyber Physical Systems and Deep Learning Chicago, Illinois 5–7 November 2018
11. Kipf, T.N., Welling, M.: Semi-supervised classification with graph convolutional networks (2016)
12. Maturana, D., Scherer, S.: Voxnet: a 3D convolutional neural network for real-time object recognition. In: 2015 IEEE/RSJ International Conference on Intelligent Robots and Systems (IROS), pp. 922–928 (2015)
13. Mehta, R., Sivaswamy, J.: M-net: a convolutional neural network for deep brain structure segmentation. In: 2017 IEEE 14th International Symposium on Biomedical Imaging (ISBI 2017), pp. 437–440 (2017)
14. Mueller, S., et al.: The Alzheimer's disease neuroimaging initiative. Clin. North Am. **15**(869–877), xi–xii (2005)
15. O'Bryant, S.E., et al.: for the HABLE Study Team: The health & aging brain among latino elders (hable) study methods and participant characteristics. Alzheimer's Dement. Diagn. Assess. Dis. Monit. **13**(1), e12202 (2021)
16. Pereira-Pedro, A.S., Bruner, E.: Sulcal pattern, extension, and morphology of the precuneus in adult humans. Ann. Anat. - Anatomischer Anz. **208**, 85–93 (2016)
17. Qiu, S., et al.: Development and validation of an interpretable deep learning framework for Alzheimer's disease classification. Brain **143**(6), 1920–1933 (2020)
18. Song, T.A., et al.: Graph convolutional neural networks for Alzheimer's disease classification. In: 2019 IEEE 16th International Symposium on Biomedical Imaging (ISBI 2019), pp. 414–417 (2019)
19. Thompson, P.M., et al.: Cortical change in Alzheimer's disease detected with a disease-specific population-based brain atlas. Cereb. Cortex **11**(1), 1–16 (2001)
20. Wang, Y., Sun, Y., Liu, Z., Sarma, S.E., Bronstein, M.M., Solomon, J.M.: Dynamic graph cnn for learning on point clouds (2018)
21. Zeiler, M.D., Fergus, R.: Visualizing and understanding convolutional networks. In: Fleet, D., Pajdla, T., Schiele, B., Tuytelaars, T. (eds.) ECCV 2014. LNCS, vol. 8689, pp. 818–833. Springer, Cham (2014). https://doi.org/10.1007/978-3-319-10590-1_53

Detection of Abnormal Folding Patterns with Unsupervised Deep Generative Models

Louise Guillon[1]([⊠])(iD), Bastien Cagna[1](iD), Benoit Dufumier[1,2](iD), Joël Chavas[1](iD), Denis Rivière[1](iD), and Jean-François Mangin[1](iD)

[1] Université Paris-Saclay, CEA, CNRS, NeuroSpin, Baobab, Gif-sur-Yvette, France
`louise.guillon@cea.fr`
[2] LTCI, Télécom Paris, IPParis, Palaiseau, France

Abstract. Although the main structures of cortical folding are present in each human brain, the folding pattern is unique to each individual. Because of this large normal variability, the identification of abnormal patterns associated to developmental disorders is a complex open challenge. In this paper, we tackle this problem as an anomaly detection task and explore the potential of deep generative models using benchmarks made up of synthetic anomalies. To focus learning on the folding geometry, brain MRI are preprocessed first to deal only with a skeleton-based negative cast of the cortex. A variational auto-encoder is trained to get a representation of the regional variability of the folding pattern of the general population. Then several synthetic benchmark datasets of abnormalities are designed. The latent space expressivity is assessed through classification experiments between control's and abnormal's latent codes. Finally, the properties encoded in the latent space are analyzed through perturbation of specific latent dimensions and observation of the resulting modification of the reconstructed images. The results have shown that the latent representation is rich enough to distinguish subtle differences like asymmetries between the right and left hemispheres.

Keywords: Variational autoencoder · Brain architecture · Cortical folding · Anomaly benchmark · Anomaly detection

1 Introduction

The cortex folds *in utero* to form numerous furrows called sulci, which delimit circumvolutions. Cortical folding is related to cortical architecture (architectony and connectivity) [7,8] and can be impacted by developmental issues that lead to brain disorders [9,25]. The identification of folding patterns acting as markers of developmental brain diseases would be a major breakthrough facilitating early diagnosis. However, although cortical morphology embeds a topography of the sulci sufficiently consistent across subjects to enable the design of automatic

A. Abdulkadir et al. (Eds.): MLCN 2021, LNCS 13001, pp. 63–72, 2021.
https://doi.org/10.1007/978-3-030-87586-2_7

recognition tools [5], the shapes of the sulci present a high diversity, which hinders the modelling of the inter-individual variability necessary to define abnormalities [18]. Hence, the diversity of the folding pattern is often put aside and canceled out using spatial normalisation, namely warping all brains toward a template space.

Associations between folding patterns and developmental disorders have already been described. For instance, a very rare pattern called the Power Button Sign (PBS) has been linked to the epileptogenic zone of patients suffering from drug-resistant type 2 focal cortical dysplasia [17]. Similarly, it was demonstrated that the paracingulate sulcus morphology is correlated to hallucinations in patients suffering from schizophrenia [24]. Sulci shape deviations have also been observed in autism spectrum disorder (ASD) [2,11].

Once abnormal folding patterns linked to a pathology have been identified, automatic detection techniques can be developed using supervised learning. For instance, the PBS can be detected with a supervised classifier [4]. However, the upstream process of identifying and defining new patterns of interest is tedious and difficult as each individual has a unique cortical folding geometry and spotting a recurrent abnormality throughout a set of patients is very complex. An unsupervised tool designed to uncover cortical folding abnormalities and potential biomarkers would be an important lever to harvest the potential meaning of unusual folding patterns.

In this paper we propose a dedicated framework based on deep learning and we test its potential through the detection of synthetic unusual folding patterns. The automatic detection of abnormal folding patterns is a challenging task that has not yet been addressed in the field of neuroimaging. In this work, we relate this objective with the general field of anomaly detection. Anomaly and novelty detection aims at identifying samples that do not fit the normal data distribution [19]. A few years ago, anomaly detection methods evolved towards deep learning approaches and specifically unsupervised deep learning due to the ability to detect potentially unseen events. Auto-encoder (AE) based methods have been particularly studied as they infer a latent space of interest with much fewer dimensions than the input space, enforcing to learn only the most common features of the training data. There exists a broad range of AE-based models. An extensive review on the detection of epilepsy lesions in MRI patches can be found in [1]. Similarly, different methods dedicated to medical images have been compared in [3] leading to qualify the variational AE (VAE) architecture as the most efficient. Generative adversarial networks (GAN) have also been used in order to identify biomarkers in optical coherence tomography scans, reaching good performances [21,22]. Based on these initial results, a first framework was proposed for anomaly detection in computed tomography scans of 3D brains where anomalies consisted in labeled traumatic brain injuries [23]. β-VAE have also been successfully used to model the inter-individual variability in the mouse brain [14]. More recently, very promising self-supervised methods have been applied to anomaly detection problems in medical images [6], but these methods lack the generative aspect provided by GAN and VAE, which is crucial in terms of explainability. All these works assessed images containing known lesions. Our aim however is to discover still unknown patterns linked to diseases, which leads to challenging evaluation issues. Therefore, this paper is focused on dedicated synthetic benchmark datasets.

In this paper, we propose the first technique aiming at bringing to light unusual and potentially abnormal folding patterns. For this purpose, we propose first a dedicated preprocessing leading to focus the learning on the cortical folding geometry of a specific region of interest (ROI). Then, like in [14], a $\beta - VAE$ is trained on a set of control data sampling the general population to get a latent representation of the folding pattern distribution in this ROI. We also create several benchmark datasets simulating unusual regional folding patterns to assess the ability of our model to detect them. Finally, we analyse the latent space capacity to separate regional patterns from the two hemispheres.

2   Methods

2.1   Focusing on Folding Information

Brain MRIs contain diverse information that are not all relevant to study folding patterns. Our method therefore includes a crucial first step of data pre-processing based on the BrainVisa/Morphologist pipeline (http://brainvisa.info) [16]. This pipeline combines several steps such as bias correction, grey-white segmentation, and skeletonization to obtain a negative cast of the folding. Morphologist's skeletons were used as input of our learning model. They were first defined in [15] and are obtained by skeletonization of the grey matter and cerebrospinal fluid union while preserving the topology. The result is 3D volumes with three values: inside of the brain, sulci skeletons and outside of the brain. The use of these simple images rather than raw MRIs puts the focus of learning on the folding geometry and discard a major confound related to the width of the sulci, which increases with local atrophies induced by aging or degenerative pathologies.

2.2   Generating Synthetic Brain Anomalies

One of our biggest challenges is the lack of consensual datasets of abnormalities to assess the approach. The examples mentioned in the introduction are either especially challenging in terms of shape (the PBS in epilepsy) or inter-subject local variability (Broca's area or Superior Temporal Sulcus (STS) branches in autism), or correspond to a stratification of the population into several frequent patterns, which is not in the scope of anomaly detection (paracingulate sulcus). Therefore, in this paper, we focus on a 3D ROI of $23 \times 37 \times 36$ voxels with 2 mm isotropic resolution, localized in only one of these challenging areas, the STS branches. This ROI has been defined in each subject using affine normalisation to the classical MNI reference space. The localization of the ROI in the MNI space has been learned from the open access training dataset with annotated sulci of the Morphologist pipeline [5]. We have designed several dedicated synthetic anomaly benchmark datasets in order to be able to evaluate the performances of our model (Fig. 1).

Deletion: Our first benchmark dataset consists of skeletons in which we have randomly deleted one piece of sulcus, which is chosen among topologically elementary parts called simple surfaces and proposed by the Morphologist pipeline [15]. To be deleted, a simple surface must be completely within the ROI and made up of more than 1000 voxels, which corresponds to about 17% of the average number of skeleton voxels in the ROI. This arbitrary threshold aims at performing modification of the geometry beyond the normal anatomical variability observed in the population.

Random: A second benchmark dataset is composed of random ROIs of the same dimensions and overlapping the skeleton but localized in different positions in the cortex. This benchmark is expected to be very easily spotted as abnormal since the images are highly different. It is used to ensure that the model is able to identify inputs that are far away from the normal distribution and that what the model has learned is not only non-region specific features such as voxel proportion and sulci continuity.

Asymmetry: The last benchmark dataset corresponds to the same ROI but defined in the other hemisphere and flipped. The flip is defined from the inter-hemispheric plane of the MNI space after affine spatial normalisation. This benchmark has a biological interest as hemispheric asymmetry is still an intense field of research.

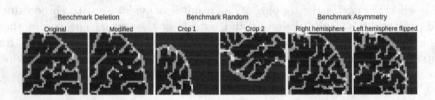

Fig. 1. *Generated anomaly benchmarks.* Benchmark Deletion: Original crop and its modified version. Benchmark Random: two examples of random crops. Benchmark Asymmetry: crop of right hemisphere and crop of left hemisphere flipped.

2.3 Learning a Representation of the Normal Variability

An effective way to model population variability is through AE-based networks. These architectures learn to project input data onto a lower dimensional manifold, also called latent space and to reconstruct from this space the input image. Simple AE are known to have some drawbacks and particularly the lack of regularization of the latent space. To overcome this issue, VAE model was introduced [13], and later $\beta - VAE$ was proposed [10]. Like classical AE, $\beta - VAE$ are composed of two parts: an encoder and a decoder but add a variational objective. Contrary to simple AE, an input from image space $\mathcal{X}$, is encoded as a distribution in a latent space $\mathcal{Z}$ comprising L dimensions, leading to a twofold objective. First, the minimization of the reconstruction error of the input image. Second, the matching of the encoded distribution to a prior distribution, usually a Gaussian, which is done thanks to Kullback-Leibler divergence and enables to regularize the latent space. VAE is a $\beta - VAE$ with KL divergence weighted at 1. Thus, $\beta - VAE$ encoder θ and decoder ϕ are trained maximising the following objective:

$$\mathcal{L}(\theta, \phi; \mathbf{x}, \mathbf{z}, \beta) = \mathbb{E}_{q_\phi(\mathbf{z}|\mathbf{x})}[\log p_\theta(\mathbf{x}|\mathbf{z})] - \beta \mathcal{D}_{KL}(q_\phi(\mathbf{z}|\mathbf{x})||p(\mathbf{z})) \qquad (1)$$

where $p(\mathbf{z})$ is the prior distribution, a reduced centered Gaussian distribution in our work that is approximated with the posterior distribution $q_\phi(\mathbf{z}|\mathbf{x})$. Tuning β parameter enables to improve latent factors disentanglement [10].

Analysing the Latent Space. The analysis of the latent space to understand the meaning of the encoded features is capital to assess the potential of our model to highlight unusual folding patterns. As such, we first trained a $\beta - VAE$ on normal data only, for our model to learn to encode normal variability. Next, normal and benchmark data unseen during training are projected to the latent space to perform this analysis. The resulting latent codes are used to train a gradient boosting algorithm to classify normal versus synthetic abnormal samples. These three classifiers are used first to ensure that the latent representations are able to capture some relevant information regarding the folding patterns. Then, we can focus our analysis on the features contributing the most to the success of the classification using the generative power of the $\beta - VAE$, like in [14]. We travel throughout the latent space modifying only one of these features and observe the generated folding patterns.

3   Results

3.1   Datasets and Implementation

To learn the inter-individual variability of control subjects, the HCP database was used[1]. MRIs were obtained with a single Siemens Skyra Connectom scanner with a resolution of 0.7 mm $\times$ 0.7 mm $\times$ 0.7 mm. In our work, we studied only the right hemisphere of 997 right-handed subjects with high quality result of the Morphologist pipeline. 547 subjects were used for the training and 150 for the validation of the $\beta - VAE$. The remaining 300 subjects were used to train classifiers, half of them being used to create synthetic abnormal patterns following each of the benchmark experiment. Two third of these 300 subjects were used to estimate the $\beta - VAE$ hyper-parameters using a grid-search driven by a classifier and the last third was used to explore the latent space organization.

The skeletons were spatially normalized with an affine transformation to the standard MNI space and were down-sampled to a voxel size of 2 mm, which

---

[1] https://www.humanconnectome.org/.

is sufficient to preserve the folding geometry. Voxels were set to 0 for inside the brain (26% of voxels), 1 for outside the brain (65%) and 2 for the sulci skeleton (9%). As mentioned above, the input to the $\beta - VAE$ was a 3D ROI, whose bounding box in MNI space was learned from the BrainVISA open access training set of 64 annoted brains, in order to include the two posterior branches of the STS. This 3D ROI is made up by $23 \times 37 \times 36$ voxels extended to $40 \times 40 \times 40$ using 1-padding.

The complete pipeline is shown in Fig. 2. Our $\beta - VAE$ was composed of fully convolutional encoder and decoder with symmetrical architectures comprising three convolutional blocks and 2 fully connected layers. We did a gridsearch (L = 8–100, β = 1–20, ranges are based on previous works [14] and reconstruction abilities), where hyperparameters were chosen according to the classification performances of the Deletion classifier applied to 100 controls and 100 synthetic samples using a 5-fold stratified cross-validation. We selected L = 100, β = 2 and a learning rate of 2e-4. Training lasted for 400 epochs on a Nvidia Quadro RTX 8000 GPU and was completed in roughly 1 h.

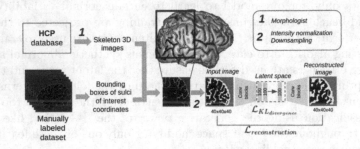

Fig. 2. *Whole pipeline.* First, bounding boxes of sulci of interest are defined. The HCP database is processed with the Morphologist pipeline and cropped thanks to the bounding boxes. Crops are then downsampled before feeding the $\beta - VAE$.

3.2 Analysing Learned Folding Variability

First, we visually evaluated the reconstruction ability of our β-VAE. 2D slices of several reconstructed inputs are presented in Fig. 3. For the control set, deletion and asymmetry benchmarks, the reconstructions are approximate but retain the general geometry. The main sulci included in the ROI can be identified. In return, the random crops cannot be reconstructed by the decoder, which outputs an image looking like a disturbed configuration of STS branches. This suggests that the model has really learned the distribution of the specific geometry of this cortical region rather than a generic distribution covering any skeleton configuration.

To evaluate our model latent space, three gradient boosting classifiers were trained using 50 controls and 50 synthetic samples from the test set, using a 5-fold stratified cross-validation. ROC curves are presented in Fig. 4.A). For

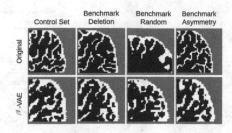

Fig. 3. *Reconstructions of test inputs.* 2D sections from the 3D ROI presented in sagittal view at depth 18. First row: original images, second row: model outputs.

each configuration, the AUC score is above chance. Highest performances are obtained on benchmark Random (AUC = 0.98), which was expected. This first result comforts the fact that our model is able to capture very obvious abnormalities. On benchmark Asymmetry, very good scores are also obtained (AUC = 0.85). Though for an inexperienced eye the difference between right hemisphere and flipped left hemisphere can seem subtle, this region is known to be asymmetrical in terms of length and tilt of the main sulci [20]. It confirms that our model is capable of representing specific anatomical structures included in the ROI. Finally, when deleting one large simple surface, results are slightly above chance (AUC = 0.69) which indicates a potential of the latent space to detect such anomalies. Nevertheless further work is required using a larger test set to overcome potential limitations of this classifier experiment.

Using gradient boosting classifiers gives insights on the most decisive dimensions of the latent space, which depend on the benchmark. Figure 4.B) shows a visualization of the datasets using the two most important latent features or tSNE 2D manifold projection. For Random and Asymmetry benchmarks, two groups can be clearly identified. Surprisingly, in the tSNE visualization, the random crops are surrounded by controls, which is counter-intuitive relative to the Gaussian prior and does not fit with the plot performed from the two most discrimative dimensions. Further experiments with unbalanced datasets including only a small ratio of abnormality will help us clarify this point. The Suppression benchmark is clearly the most difficult classification experiment with the current latent space.

The last experiment consists in sampling latent vectors to travel along the most important dimensions according to the classifiers. Figure 4.C) shows the reconstruction provided by the decoder when following the dimension corresponding to the best discriminator of Asymmetry benchmark. All other dimensions were fixed at their mean value. As expected, the outputs look like approximated skeletons of the ROI. Subtle changes consistent during the travel can be observed: i.e. the upper part of the sulcus presented in pink on Fig. 4.C), called the Sylvian fissure (SF), seems to shorten from the left to the right of the dimension. This observation is consistent with Asymmetry benchmark's subjects distribution according to the two most important features on Fig. 4.B). Indeed,

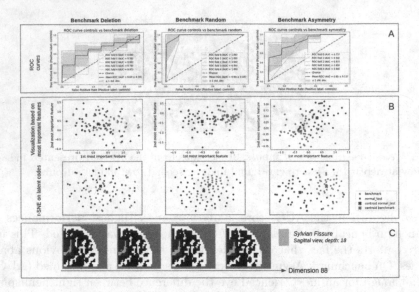

Fig. 4. *Analyses of latent space.* **A)** ROC curves of GradientBoosting classifiers for classifications between controls and benchmarks. **B)** Distribution of data points using classifiers most important features and t-SNE. **C)** Travelling through the latent space.

control subjects, i.e. right hemispheres correspond to higher values of the 88th dimension which is also suggested by the SF shortening on Fig. 4.C). This evolution is interesting as previous works demonstrated that the SF was shorter in the right hemisphere. Thus, this dimension could encode the length of the SF which is an important marker of laterality [12].

4   Discussion and Conclusion

In this paper we developed a framework that shall lead to the discovery of abnormal folding patterns beyond reach for human cognition because of the high inter-subject normal variability. Our main contribution is the design of synthetic benchmarks used to decipher the organization of the latent space used to model this variability. We have shown that the regional specificity of the folding pattern can be learned and used to detect some deviations from the norm. Our methods achieved to detect obvious and more subtle deviations (respectively with Random and Asymmetry benchmarks), but detection is harder for more complex ones such as Deletion benchmark. However, we don't seek to detect benchmark subjects but rather to detect abnormal patterns. In the future we plan to perform further experiments to get more hints about the nature of the representation. Although some interesting features can be observed when varying latent vectors dimensions' values, it is very difficult to apprehend all possible shifts looking at 2D sections because we deal with 3D images. Additionally, in some cases, we observe that the generated images are not realistic, preventing the interpretation. Generating synthetic anomaly benchmarks is a very useful initial step but

induces biases that must be acknowledged. In our case, with the Deletion benchmark we made the hypothesis that anomalies could be linked to "missing" simple
surfaces. This work constitutes a first step of proof of concept and enables to
control the complexity of abnormalities, however an important limitation is the
high dependency of the results on simulated data as our benchmarks are made of
synthetic anomalies. Nevertheless we stress out that the model has learned only
control data, thus is totally unsupervised in regards to anomaly data, and only
the evaluation results depend on simulated data. In future works, the ultimate
benchmarks will have to be built from large datasets of neurodevelopmental
disorders and will aim at discovering actual abnormal folding patterns.

Acknowledgments. This project has received funding from the FRMDIC2016123
6445, the ANR-19-CE45-0022-01 IFOPASUBA, the ANR-14-CE30-0014-02 APEX the
ANR-20-CHIA-0027-01 FOLDDICO. Data were provided in part by the Human Connectome Project funded by the NIH. This work was performed using HPC resources
from GENCI-IDRIS (Grant 2020-AD011011929).

References

1. Alaverdyan, Z.: Unsupervised representation learning for anomaly detection on
   neuroimaging. Application to epilepsy lesion detection on brain MRI. Ph.D. thesis,
   Université de Lyon (2019)
2. Auzias, G., et al.: Atypical sulcal anatomy in young children with autism spectrum
   disorder. NeuroImage: Clin. **4**, 593–603 (2014). https://doi.org/10.1016/j.nicl.2014.
   03.008
3. Baur, C., Denner, S., Wiestler, B., Albarqouni, S., Navab, N.: Autoencoders for
   Unsupervised Anomaly Segmentation in Brain MR Images: A Comparative Study.
   arXiv:2004.03271 [cs, eess] (2020)
4. Borne, L., et al.: Automatic recognition of specific local cortical folding patterns.
   NeuroImage **238** (2021). https://doi.org/10.1016/j.neuroImage.2021.118208
5. Borne, L., Rivière, D., Mancip, M., Mangin, J.-F.: Automatic labeling of cortical
   sulci using patch- or CNN-based segmentation techniques combined with bottom-
   up geometric constraints. Med. Image Anal. **62** (2020). https://doi.org/10.1016/j.
   media.2020.101651
6. Bozorgtabar, B., Mahapatra, D., Vray, G., Thiran, J.P.: SALAD: self-supervised
   aggregation learning for anomaly detection on X-Rays. In: Martel, A.L., et al.
   (eds.) Medical Image Computing and Computer Assisted Intervention - MICCAI
   2020, pp. 468–478. Lecture Notes in Computer Science, Springer International
   Publishing, Cham (2020). https://doi.org/10.1007/978-3-030-59710-8_46
7. Fernandez, V., Llinares-Benadero, C., Borrell, V.: Cerebral cortex expansion and
   folding: what have we learned? EMBO J. **35**(10), 1021–1044 (2016). https://doi.
   org/10.15252/embj.201593701
8. Fischl, B., et al.: Cortical folding patterns and predicting cytoarchitecture. Cereb.
   Cortex **18**(8), 1973–1980 (2008). https://doi.org/10.1093/cercor/bhm225
9. Guerrini, R., Dobyns, W.B., Barkovich, A.J.: Abnormal development of the human
   cerebral cortex: genetics, functional consequences and treatment options. Trends
   Neurosci. **31**(3), 154–162 (2008). https://doi.org/10.1016/j.tins.2007.12.004
10. Higgins, I., et al.: beta-VAE: Learning Basic Visual Concepts with a Constrained
    Variational Framework (2016). https://openreview.net/forum?id=Sy2fzU9gl

11. Hotier, S., et al.: Social cognition in autism is associated with the neurodevelopment of the posterior superior temporal sulcus. Acta Psychiatr. Scand. **136**(5), 517–525 (2017). https://doi.org/10.1111/acps.12814

12. Idowu, O.E., Soyemi, S., Atobatele, K.: Morphometry, asymmetry and variations of the Sylvian fissure and sulci bordering and within the pars triangularis and pars operculum: an autopsy study. J. Clin. Diagn. Res. JCDR **8**(11), AC11–AC14 (2014). https://doi.org/10.7860/JCDR/2014/9955.5130

13. Kingma, D.P., Welling, M.: Auto-Encoding Variational Bayes [cs, stat] May 2014. arXiv:1312.6114

14. Liu, R., et al.: A generative modeling approach for interpreting population-level variability in brain structure. In: Martel, A.L., et al. (eds.) MICCAI 2020. LNCS, vol. 12265, pp. 257–266. Springer, Cham (2020). https://doi.org/10.1007/978-3-030-59722-1_25

15. Mangin, J.F., Frouin, V., Bloch, I., Rigis, J., Lopez-Krahe, J.: From 3D magnetic resonance images to structural representations of the cortex topography using topology preserving deformations. J. Math. Imaging Vis. **5**(4), 297–318 (1995)

16. Mangin, J.F., et al.: Object-based morphometry of the cerebral cortex. IEEE Trans. Med. Imaging **23**, 968–82 (2004). https://doi.org/10.1109/TMI.2004.831204. Sep

17. Mellerio, C., et al.: The power button sign: a newly described central sulcal pattern on surface rendering MR images of type 2 focal cortical dysplasia. Radiology **274**(2), 500–507 (2014). https://doi.org/10.1148/radiol.14140773, publisher: Radiological Society of North America

18. Ono, M., Kubik, S., Abernathey, C.D.: Atlas of the cerebral sulci. G. Thieme Verlag; Thieme Medical Publishers, Stuttgart; New York (1990). oCLC: 645306373

19. Pang, G., Shen, C., Cao, L., Hengel, A.V.D.: Deep Learning for Anomaly Detection: A Review. arXiv:2007.02500 [cs, stat] July 2020

20. Rubens, A.B., Mahowald, M.W., Hutton, J.T.: Asymmetry of the lateral (Sylvian) fissures in man. Neurology **26**(7), 620–620 (1976)

21. Schlegl, T., Seeböck, P., Waldstein, S.M., Langs, G., Schmidt-Erfurth, U.: f-AnoGAN: fast unsupervised anomaly detection with generative adversarial networks. Med. Image Anal. **54**, 30–44 (2019). https://doi.org/10.1016/j.media.2019.01.010

22. Schlegl, T., Seeböck, P., Waldstein, S.M., Schmidt-Erfurth, U., Langs, G.: Unsupervised Anomaly Detection with Generative Adversarial Networks to Guide Marker Discovery. arXiv:1703.05921 [cs] March 2017

23. Simarro Viana, J., de la Rosa, E., Vande Vyvere, T., Robben, D., Sima, D.M., Investigators, C.E.N.T.E.R.-T.B.I.P.: Unsupervised 3D brain anomaly detection. In: Crimi, A., Bakas, S. (eds.) BrainLes 2020. LNCS, vol. 12658, pp. 133–142. Springer, Cham (2021). https://doi.org/10.1007/978-3-030-72084-1_13

24. The Australian Schizophrenia Research Bank, Garrison, J.R., Fernyhough, C., McCarthy-Jones, S., Haggard, M., Simons, J.S.: Paracingulate sulcus morphology is associated with hallucinations in the human brain. Nat. Commun. **6**(1), 8956 (2015). https://doi.org/10.1038/ncomms9956

25. Walsh, C.A.: Genetic malformations of the human cerebral cortex. Neuron **23**(1), 19–29 (1999). https://doi.org/10.1016/S0896-6273(00)80749-7

PialNN: A Fast Deep Learning Framework for Cortical Pial Surface Reconstruction

Qiang Ma[1]([✉]), Emma C. Robinson[2], Bernhard Kainz[1], Daniel Rueckert[1], and Amir Alansary[1]

[1] BioMedIA, Department of Computing, Imperial College London, London, UK
q.ma20@imperial.ac.uk
[2] School of Biomedical Engineering and Imaging Sciences, King's College London, London, UK

Abstract. Traditional cortical surface reconstruction is time consuming and limited by the resolution of brain Magnetic Resonance Imaging (MRI). In this work, we introduce Pial Neural Network (PialNN), a 3D deep learning framework for pial surface reconstruction. PialNN is trained end-to-end to deform an initial white matter surface to a target pial surface by a sequence of learned deformation blocks. A local convolutional operation is incorporated in each block to capture the multi-scale MRI information of each vertex and its neighborhood. This is fast and memory-efficient, which allows reconstructing a pial surface mesh with 150k vertices in one second. The performance is evaluated on the Human Connectome Project (HCP) dataset including T1-weighted MRI scans of 300 subjects. The experimental results demonstrate that PialNN reduces the geometric error of the predicted pial surface by 30% compared to state-of-the-art deep learning approaches. The codes are publicly available at https://github.com/m-qiang/PialNN.

1 Introduction

As an essential part in neuroimage processing, cortical surface reconstruction aims to extract 3D meshes of the inner and outer surfaces of the cerebral cortex from brain MRI, also known as the white matter and pial surfaces. The extracted surface can be further analyzed for the prediction and diagnosis of brain diseases as well as for the visualisation of information on the cortex. However, it is difficult to extract a geometrically accurate and topologically correct cortical surface due to its highly curved and folded geometric shape [3,6].

The typical cortical surface reconstruction pipeline, which can be found in existing neuroimage analysis tools [1,3,5,8,15], consists of two main steps. Firstly, an initial white matter surface mesh is created by applying mesh tessellation or marching cubes [12] to the segmented white matter from the scanned image, along with topology fixing to guarantee the spherical topology. The initial mesh is further refined and smoothed to produce the final white matter surface.

© Springer Nature Switzerland AG 2021
A. Abdulkadir et al. (Eds.): MLCN 2021, LNCS 13001, pp. 73–81, 2021.
https://doi.org/10.1007/978-3-030-87586-2_8

Secondly, the pial surface mesh is generated by expanding the white matter surface iteratively until it reaches the boundary between the gray matter and cerebrospinal fluid or causes self-intersection. One limitation of such approaches is the high computational cost. For example, FreeSurfer [5], a widely used brain MRI analysis tool, usually takes several hours to extract the cortical surfaces for a single subject.

As a fast and end-to-end alternative approach, deep learning has shown its advantages in surface reconstruction for general shape objects [7,9,13,14,18] and medical images [2,10,16,19]. Given brain MRI scans, existing deep learning frameworks [2,10] are able to predict cortical surfaces within 30 min. However, although the white matter surfaces can be extracted accurately, the pial surface reconstruction is still challenging. Due to its highly folded and curved geometry, the pial surface reconstructed by previous deep learning approaches tends to be oversmooth to prevent self-intersections, or fails to reconstruct the deep and narrow sulcus region.

In this work, we propose a fast and accurate architecture for reconstructing the pial surface, called Pial Neural Network (PialNN). Given an input white matter surface and its corresponding MR image, PialNN reconstructs the pial surface mesh using a sequence of learned deformation blocks. In each block, we introduce a local convolutional operation, which applies a 3D convolutional neural network (CNN) to a small cube containing the MRI intensity of a vertex and its neighborhood. Our method can work on brain MRI at arbitrary resolution without increasing the complexity. PialNN establishes a one-to-one correspondence between the vertices in white matter and pial surface, so that a point-to-point loss can be minimized directly without any regularization terms or point matching. The performance is evaluated on the publicly available Human Connectome Project (HCP) dataset [17]. PialNN shows superior geometric accuracy compared to existing deep learning approaches.

The main contributions and advantages of PialNN can be summarized as:

- **Fast:** PialNN can be trained end-to-end to reconstruct the pial surface mesh within one second.
- **Memory-efficient:** The local convolutional operation enables PialNN to process a high resolution input mesh (>150k vertices) using input MR brain images at arbitrary resolution.
- **Accurate:** The proposed point-to-point loss, without additional vertex matching or mesh regularization, improves the geometric accuracy of the reconstructed surfaces effectively.

2   Related Work

Deep learning-based surface reconstruction approaches can be divided into implicit [13,14] and explicit methods [7,9,18]. The former use a deep neural network (DNN) to learn an implicit surface representation such as an occupancy field [13] and a signed distance function [14]. A triangular mesh is then extracted using isosurface extraction. For explicit methods [7,9,18], a DNN is trained

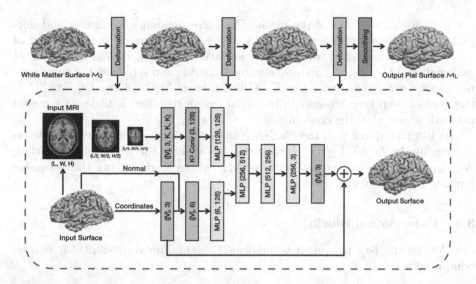

Fig. 1. The proposed architecture for pial surface reconstruction (PialNN). The input white matter surface is deformed by three deformation blocks to predict a pial surface. Each deformation block incorporates two types of features: point features from the white matter surface vertices and local features from the brain MRI. Finally, the output mesh is refined using Laplacian smoothing.

end-to-end to deform an initial mesh to a target mesh, producing an explicit mesh directly.

Previous deep learning frameworks [2,10] for cortical surface reconstruction mainly adopted implicit methods. Henschel et al. [10] proposed FastSurfer pipeline, which improved FreeSurfer [5] by introducing a fast CNN for whole brain segmentation instead of atlas-based registration. The cortical surface is then extracted by a non-learning approach [5]. Cruz et al. [2] proposed DeepCSR framework to predict the implicit representation of both the inner and outer cortical surfaces. Explicit surfaces are extracted by the marching cubes algorithm [12]. Implicit methods require a time-consuming topology correction, while explicit methods can pre-define an initial mesh with spherical topology to achieve fast inference. Wickramasinghe et al. [19] presented an explicit framework, called Voxel2Mesh, to extract 3D meshes from medical images. Voxel2Mesh employed a series of deformation and unpooling layers to deform an initial mesh while increasing the number of vertices iteratively. Regularization terms are utilized to improve the mesh quality and prevent self-intersections, whereas these terms tend to oversmooth the output mesh. Conversely, our PialNN uses explicit methods to learn the pial surface reconstruction without any regularization terms.

3   Method

We first introduce necessary notations to formulate the problem. Let $\mathcal{M} = (\mathcal{V}, \mathcal{E}, \mathcal{F})$ be a 3D triangular mesh, where $\mathcal{V} \subset \mathbb{R}^3$, $\mathcal{E}$ and $\mathcal{F}$ are the sets of vertices, edges and faces of the mesh. The corresponding coordinates and normal of the vertices are represented by $\mathbf{v}, \mathbf{n} \in \mathbb{R}^{|\mathcal{V}| \times 3}$, where $|\mathcal{V}|$ is the number of vertices. Given an initial white matter surface $\mathcal{M}_0 = (\mathcal{V}_0, \mathcal{E}_0, \mathcal{F}_0)$ and a target pial surface $\mathcal{M}_* = (\mathcal{V}_*, \mathcal{E}_*, \mathcal{F}_*)$, we assume that $\mathcal{M}_0$ and $\mathcal{M}_*$ have the same connectivity, i.e. $\mathcal{E}_0 = \mathcal{E}_*$ and $\mathcal{F}_0 = \mathcal{F}_*$. Given a brain MRI volume $\mathbf{I} \in \mathbb{R}^{L \times W \times H}$, the goal of deep learning-based pial surface reconstruction is to learn a neural network g such that the coordinates $\mathbf{v}_* = g(\mathbf{v}_0, \mathbf{n}_0, \mathbf{I})$.

As illustrated in Fig. 1, the PialNN framework aims to learn a series of deformation blocks f_{θ_l} for $1 \leq l \leq L$ to iteratively deform the white matter surface $\mathcal{M}_0$ to match the target pial surface $\mathcal{M}_*$, where θ_l represents the learnable parameter of the neural network.

3.1 Deformation Block

Let $\mathcal{M}_l$ be the l-th intermediate deformed mesh. The vertices of $\mathcal{M}_l$ can be computed as:

$$\mathbf{v}_l = \mathbf{v}_{l-1} + \Delta \mathbf{v}_{l-1} = \mathbf{v}_{l-1} + f_{\theta_l}(\mathbf{v}_{l-1}, \mathbf{n}_{l-1}, \mathbf{I}), \tag{1}$$

for $1 \leq l \leq L$, where f_{θ_l} is the l-th deformation block represented by a neural network. The purpose of PialNN is to learn the optimal f_{θ_l}, such that the final predicted mesh $\mathcal{M}_L$ matches the target mesh $\mathcal{M}_*$, i.e. $\mathcal{M}_L = \mathcal{M}_*$. The architecture of the deformation block is shown in Fig. 1. In this approach, the deformation block predicts a displacement $\Delta \mathbf{v}$ based on the *point feature* and *local feature* of the vertex $\mathbf{v}$.

Point Feature. The point feature of a vertex is defined as the feature extracted from its coordinate $\mathbf{v}$ and normal $\mathbf{n}$, which includes the spacial and orientation information. We extract the point feature using a multi-layer perceptron (MLP).

Local Feature. We adopt a local convolutional operation to extract the local feature of a vertex from brain MRI scans. Rather than using a memory-intensive 3D CNN on the entire MRI volume [19], this method only employs a CNN on a cube containing MRI intensity of each vertex and its neighborhood. As illustrated in Fig. 1, for each vertex, we find the corresponding position in the brain MRI volume. Then a K^3 grid is constructed based on the vertex to exploit its neighborhood information. The voxel value of each point in the grid is sampled from the MRI volume. Such a cube sampling approach extracts a K^3 voxel cube containing the MRI intensity of each vertex and its neighborhood.

Furthermore, we build a 3D image pyramid including 3 scales (1, 1/2, 1/4) and use cube sampling on the different scales. Therefore, each vertex is represented by a K^3 local cube with 3 channels containing multi-scale information. A 3D CNN with kernel size K is then applied to each local cube, which converts the cube to a local feature vector of its corresponding vertex. An MLP layer is followed to further refine the local feature.

Such a local convolutional operation is memory- and time-efficient. As there are total $|\mathcal{V}|$ cubes with 3 channels, it only executes the convolution operators

$3|\mathcal{V}|$ times, which are far less than $L \times W \times H$ times for running a 3D CNN on the full MRI. Since the complexity only relies on the number of vertices $|\mathcal{V}|$, the local convolutional operation can process MRI volumes at arbitrary resolution without increasing the complexity.

The point and local features are concatenated as the input of several MLP layers followed by leaky ReLU activation, which predict a 3D displacement $\Delta \mathbf{v}_{l-1}$. The new vertices $\mathbf{v}_l$ are updated according to Eq. 1, and act as the input for the next deformation block.

3.2   Smoothing and Training

Laplacian Smoothing. After three deformation blocks, a Laplacian smoothing is used to further smooth the surface and prevent self-intersections. For each vertex $v^i \in \mathbb{R}^3$, the smoothing is defined as $\bar{v}^i = (1-\lambda)v^i + \lambda \sum_{j \in \mathcal{N}(i)} v^j / |\mathcal{N}(i)|$, where λ controls the degree of smoothness and $\mathcal{N}(i)$ is the adjacency list of the i-th vertex. The smoothing layer is incorporated in both training and testing.

Loss Function. The Chamfer distance [4] is commonly used as the loss function for training explicit surface reconstruction models [18,19]. It measures the distance from a vertex in one mesh to the closest vertex in the other mesh bidirectionally. For PialNN, since the input and target mesh have the same connectivity, we can directly compute a point-to-point mean square error (MSE) loss between each pair of vertices. Therefore, the loss function is defined as:

$$\mathcal{L}(\mathcal{M}_L, \mathcal{M}_*) = \mathcal{L}(\mathbf{v}_L, \mathbf{v}_*) = \|\mathbf{v}_L - \mathbf{v}_*\|_2^2. \tag{2}$$

Rather than computing the loss for all intermediate meshes $\mathcal{M}_l$, we only compute the loss between the final predicted pial surface $\mathcal{M}_L$ and the ground truth $\mathcal{M}_*$, because the gradient can be backpropagated to all deformation blocks f_{θ_l} for $1 \leq l \leq L$. The parameters θ_l are learned by minimizing the MSE loss.

It is noted that no explicit regularization term is required in the loss function, as the vertex will learn from the point-to-point supervision to move to a correct location. Such loss function effectively improves the geometric accuracy of the output mesh. Besides, we use an additional Laplacian smoothing after training to improve the mesh quality and to fix self-intersections.

4   Experiments

Dataset. The proposed framework is evaluated using the WU-Minn Human Connectome Project (HCP) Young Adult dataset [17][1]. We use 300 subjects, each of which has T1-weighted brain MRI scans with 1 mm isotropic resolution. Each brain MRI is cropped to size of (192, 224, 192). The 300 subjects are split into 200/50/50 for training/validation/testing. The input white matter surface and ground truth pial surface are generated by FreeSurfer [5]. Each surface has approximately 150k vertices and 300k faces for one hemisphere. It is noted that the input white matter surfaces can be generated by other faster tools [10, 15].

---

[1]   https://www.humanconnectome.org/study/hcp-young-adult/data-releases

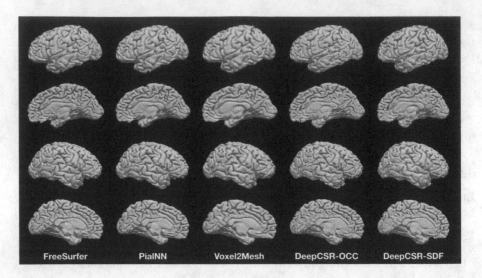

Fig. 2. Visualization of the reconstructed pial surface meshes.

Implementation Details. PialNN consists of $L = 3$ layers of deformation blocks. We set the smoothing coefficient $\lambda = 1$ and kernel size $K = 5$ for 3D CNN. The Adam optimizer with learning rate 10^{-4} is used for training the model for 200 epochs with batch size 1. Experiments compare the performance of PialNN with state-of-the-art deep learning baselines, such as Voxel2Mesh [19] and DeepCSR [2]. All models are trained on an Nvidia GeForce RTX3080 GPU.

Since Voxel2Mesh uses iterative mesh unpooling, the input white matter surface is simplified to a mesh with 5120 faces using quadric error metric decimation. For DeepCSR, we train two different models based on occupancy fields (DeepCSR-OCC) and signed distance functions (DeepCSR-SDF) for ground truth. The size of the implicit representation for DeepCSR is set to (192, 224, 192) in order to have a reasonable number of vertices for a fair comparison.

Geometric Accuracy. We evaluate the geometric accuracy of the PialNN framework by computing the error between the predicted pial surfaces and FreeSurfer ground truth. We utilize three distance-based metrics to measure the geometric error, namely, Chamfer distance (CD) [4,18], average absolute distance (AD) [2] and Hausdorff distance (HD) [2]. The CD measures the mean distance between two sets of vertices. AD and HD compute the average and maximum distance between two sets of 150k sampled points from surface meshes. All distances are computed bidirectionally in millimeters (mm). A lower distance means a better result. The experimental results are given in Table 1, which shows that PialNN achieves the best geometric accuracy compared with existing deep learning baselines. It reduces the geometric error by >30% compared to Voxel2Mesh and DeepCSR in all three distances (mm). In addition, the quality of the predicted pial surface mesh is visualized in Fig. 2.

Table 1. Geometric error for pial surface reconstruction. The results include the comparison with existing deep learning baselines and the ablation study. Chamfer distance (mm), average absolute distance (mm), and Hausdorff distance (mm) are computed for both left and right hemisphere. A lower distance means a better result.

	Left Pial			Right Pial		
Method	Chamfer	Average	Hausdorff	Chamfer	Average	Hausdorff
PialNN (Ours)	**0.39 ± 0.01**	**0.21 ± 0.02**	**0.45 ± 0.04**	**0.39 ± 0.02**	**0.20 ± 0.02**	**0.44 ± 0.04**
Voxel2Mesh	0.58 ± 0.03	0.34 ± 0.04	0.82 ± 0.09	0.57 ± 0.02	0.31 ± 0.02	0.80 ± 0.07
DeepCSR-OCC	0.66 ± 0.04	0.42 ± 0.04	0.87 ± 0.13	0.65 ± 0.05	0.40 ± 0.04	0.88 ± 0.20
DeepCSR-SDF	0.72 ± 0.07	0.45 ± 0.06	1.23 ± 0.36	0.78 ± 0.11	0.49 ± 0.09	1.58 ± 0.54
Single Scale	0.42 ± 0.02	0.23 ± 0.02	0.50 ± 0.05	0.43 ± 0.02	0.23 ± 0.02	0.51 ± 0.05
Point Sampling	0.56 ± 0.03	0.40 ± 0.04	0.87 ± 0.09	0.57 ± 0.03	0.41 ± 0.05	0.91 ± 0.11
GCN	0.39 ± 0.02	0.21 ± 0.02	0.46 ± 0.04	0.40 ± 0.01	0.21 ± 0.01	0.46 ± 0.04

Figure 3 provides a detailed visual comparison between different approaches. The DeepCSR-SDF-2x represents DeepCSR-SDF with input size of (384, 448, 384). We focus on the areas highlighted by the blocks in different colors. In the red block, the DeepCSR frameworks fail to distinguish two separate regions in the surface. The issue remains unsolved after increasing the input size. The yellow block indicates an inaccurate Voxel2Mesh prediction since the mesh is over-smoothed. In the orange block, only FreeSurfer and PialNN reconstruct the deep and narrow sulci accurately. The green block indicates the error of Voxel2Mesh and DeepCSR-OCC in a sulcus region. It is noted that PialNN makes a correct reconstruction in all highlighted areas.

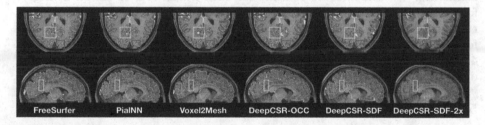

FreeSurfer   PialNN   Voxel2Mesh   DeepCSR-OCC   DeepCSR-SDF   DeepCSR-SDF-2x

Fig. 3. A visual evaluation of the predicted pial surfaces (cyan colour). (Color figure online)

Figure 3 further shows that the predicted mesh from Voxel2Mesh is over-smoothed, which can be a result of the used regularization terms. Besides, it loses the geometric prior provided by the input white matter surface due to the mesh simplification. Regardless of the input size, DeepCSR is prone to fail in the deep sulcus regions, since the implicit representation can be affected by partial volume effects.

Ablation Study. We consider three ablation experiments. First, we only use single-scale brain MRI rather than a multi-scale image pyramid. Second, instead of cube sampling, we only employ point sampling, which samples the MRI voxels at the exact position of each vertex. Third, we substitute the MLP layers with

Graph Convolutional Networks (GCN) [11]. The results are listed in Table 1 and the error maps are given in Fig. 4, which shows the Chamfer distance between the output surface and the FreeSurfer ground truth. Multi-scale input slightly improves the geometric accuracy, while the cube sampling contributes a lot to the performance of PialNN. There is no notable improvement after replacing MLP with GCN layers but the memory usage has increased.

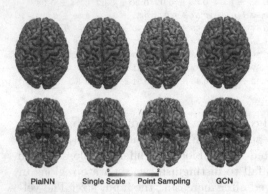

PialNN   Single Scale   Point Sampling   GCN

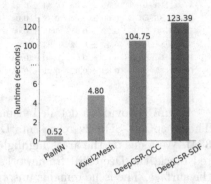

Fig. 4. Error maps of the pial surface from ablation study. The color visualizes the Chamfer distance ranging from 0 to 2 mm. (Color figure online)

Fig. 5. Runtime (seconds) of deep learning-based approaches for pial surface reconstruction.

Runtime. We compute the runtime for each framework, as shown in Fig. 5, for both left and right pial surfaces reconstruction. PialNN achieves the fastest runtime with 0.52 s, whereas traditional pipelines [5,8,15] usually take >10 min for pial surface generation based on the white matter surface. Voxel2Mesh needs 4.8 s as it requires mesh simplification for the input. DeepCSR runs in >100 s due to the time-consuming topology correction.

5   Conclusion

PialNN is a fast and memory-efficient deep learning framework for cortical pial surface reconstruction. The proposed framework learns several deformation blocks to generate a pial surface mesh from an input white matter surface. Each block incorporates the point feature extracted from the coordinates and normals, as well as the local feature extracted from the MRI intensity of the vertex and its neighborhood. Experiments demonstrate that our framework achieves the best performance with highest accuracy and fastest runtime (within one second) compared to state-of-the-art deep learning baselines. A future direction will be to extend the PialNN framework to predict the segmentation labels and reconstruct both cortical white matter and pial surfaces using only the input MR brain images.

Acknowledgements. This work was supported by the President's PhD Scholarships at Imperial College London.

References

1. Avants, B.B., Tustison, N., Song, G.: Advanced normalization tools (ANTs). Insight J. **2**(365), 1–35 (2009)
2. Cruz, R.S., Lebrat, L., Bourgeat, P., Fookes, C., Fripp, J., Salvado, O.: DeepCSR: A 3D deep learning approach for cortical surface reconstruction. arXiv preprint arXiv:2010.11423 (2020)
3. Dale, A.M., Fischl, B., Sereno, M.I.: Cortical surface-based analysis: I. segmentation and surface reconstruction. Neuroimage **9**(2), 179–194 (1999)
4. Fan, H., Su, H., Guibas, L.J.: A point set generation network for 3D object reconstruction from a single image. In: Proceedings of the IEEE Conference on Computer Vision and Pattern Recognition, pp. 605–613 (2017)
5. Fischl, B.: FreeSurfer. Neuroimage **62**(2), 774–781 (2012)
6. Fischl, B., Dale, A.M.: Measuring the thickness of the human cerebral cortex from magnetic resonance images. Proc. Natl. Acad. Sci. **97**(20), 11050–11055 (2000)
7. Gkioxari, G., Malik, J., Johnson, J.: Mesh R-CNN. In: Proceedings of the IEEE International Conference on Computer Vision, pp. 9785–9795 (2019)
8. Glasser, M.F., et al.: The minimal preprocessing pipelines for the human connectome project. Neuroimage **80**, 105–124 (2013)
9. Groueix, T., Fisher, M., Kim, V.G., Russell, B.C., Aubry, M.: A papier-mâché approach to learning 3D surface generation. In: Proceedings of the IEEE Conference on Computer Vision and Pattern rRecognition, pp. 216–224 (2018)
10. Henschel, L., Conjeti, S., Estrada, S., Diers, K., Fischl, B., Reuter, M.: FastSurfer - a fast and accurate deep learning based neuroimaging pipeline. NeuroImage **219**, 117012 (2020)
11. Kipf, T.N., Welling, M.: Semi-supervised classification with graph convolutional networks. arXiv preprint arXiv:1609.02907 (2016)
12. Lorensen, W.E., Cline, H.E.: Marching cubes: a high resolution 3D surface construction algorithm. ACM Siggraph Comput. Graph. **21**(4), 163–169 (1987)
13. Mescheder, L., Oechsle, M., Niemeyer, M., Nowozin, S., Geiger, A.: Occupancy networks: learning 3D reconstruction in function space. In: Proceedings of the IEEE Conference on Computer Vision and Pattern Recognition, pp. 4460–4470 (2019)
14. Park, J.J., Florence, P., Straub, J., Newcombe, R., Lovegrove, S.: DeepSDF: learning continuous signed distance functions for shape representation. In: Proceedings of the IEEE/CVF Conference on Computer Vision and Pattern Recognition, pp. 165–174 (2019)
15. Shattuck, D.W., Leahy, R.M.: BrainSuite: an automated cortical surface identification tool. Med. Image Anal. **6**(2), 129–142 (2002)
16. Tóthová, K., et al.: Probabilistic 3D surface reconstruction from sparse MRI information. In: Martel, A.L., et al. (eds.) MICCAI 2020. LNCS, vol. 12261, pp. 813–823. Springer, Cham (2020). https://doi.org/10.1007/978-3-030-59710-8_79
17. Van Essen, D.C., et al.: The WU-Minn human connectome project: an overview. Neuroimage **80**, 62–79 (2013)
18. Wang, N., et al.: Pixel2Mesh. 3D mesh model generation via image guided deformation. IEEE Trans. Pattern Anal. Mach. Intell. (2020)
19. Wickramasinghe, U., Remelli, E., Knott, G., Fua, P.: Voxel2Mesh: 3D mesh model generation from volumetric data. In: Martel, A.L., et al. (eds.) MICCAI 2020. LNCS, vol. 12264, pp. 299–308. Springer, Cham (2020). https://doi.org/10.1007/978-3-030-59719-1_30

Multi-modal Brain Segmentation Using Hyper-Fused Convolutional Neural Network

Wenting Duan[1(✉)], Lei Zhang[1], Jordan Colman[1,2], Giosue Gulli[2], and Xujiong Ye[1]

[1] Department of Computer Science, University of Lincoln, Lincoln, UK
wduan@lincoln.ac.uk
[2] Ashford and St Peter's Hospitals NHS Foundation Trust, Surrey, UK

Abstract. Algorithms for fusing information acquired from different imaging modalities have shown to improve the segmentation results of various applications in the medical field. Motivated by recent successes achieved using densely connected fusion networks, we propose a new fusion architecture for the purpose of 3D segmentation in multi-modal brain MRI volumes. Based on a hyper-densely connected convolutional neural network, our network features in promoting a progressive information abstraction process, introducing a new module – ResFuse to merge and normalize features from different modalities and adopting combo loss for handing data imbalances. The proposed approach is evaluated on both an outsourced dataset for acute ischemic stroke lesion segmentation and a public dataset for infant brain segmentation (iSeg-17). The experiment results show our approach achieves superior performances for both datasets compared to the state-of-art fusion network.

Keywords: Multi-modal fusion · Dense network · Brain segmentation

1 Introduction

In medical imaging, segmentation of lesions or organs using a multi-modal approach has become a growing trend strategy as more advanced systems and data becomes available. For example, magnetic resonance imaging (MRI) that is widely used for brain lesion or tumor detection and segmentation comes in several modalities including T1-weighted (T1), T2-weighted (T2), FLuid Attenuated Inversion Recovery (FLAIR) and Diffusion-weighted image (DWI), etc. Compared to single modality, the extraction of information from multi-modal images brings complementary information that contributes to reduced uncertainty and an improved discriminative power of the clinical diagnosis system [1]. Motivated by the success of deep learning, image fusion strategies have largely moved from probability theory [2] or fuzzy concept [3] based methods to deep convolutional neural network based approaches [1, 4].

Promising performance has been achieved by deep learning based methods for medical image segmentation from multi-modal images. The most widely applied strategy is simply concatenating images or image patches of different modalities to learn a unified image features set [5–7]. Such networks combine the data at the input level to

© Springer Nature Switzerland AG 2021
A. Abdulkadir et al. (Eds.): MLCN 2021, LNCS 13001, pp. 82–91, 2021.
https://doi.org/10.1007/978-3-030-87586-2_9

form a multi-channel input. Another straightforward fusion strategy is for images of each modality to learn an independent feature map. Then these single-modality feature sets will, either learn their separate classifiers and use 'votes' to arrive at a final output, or learn a multi-modal classifier integrating high-level representations of different modalities [8–10]. In comparison to the strategies mentioned previously where fusion happens either at the input level or the output/classifier level, some recent works [11–14] have proved that performing fusion within the convolutional feature learning stage instead generally gives much better segmentation results. Tseng *et al.* [14] proposed a cross-modality convolution to aggregate data from different modalities within an encoder decoder network. The convolution LSTM is then used to model the correlations between slices. The method requires images of all modalities to be co-registered and the network parameters varies with the number of slices involved in the training dataset. For unpaired modalities such as CT and MRI, Dou *et al.* [15] developed a novel scheme involving separate feature normalization but shared convolution. Knowledge distillation-based loss is proposed to promote softer probability distribution over classes. However, the design so far is limited to two modalities. Another avenue of research on multi-modal fusion is based on DenseNet [16] where feature re-use is induced by connecting each layer with all previous layers. For example, Dolz *et al.* [13] extends the DenseNet so that the dense connections not only exist in the layers of same modality but also between the modalities. Their network (i.e. HyperDense-Net) made significant improvements over other state-of-art segmentation techniques and ranked first for two highly competitive multi-modal brain segmentation challenges. Dolz *et al.* [17] also explored the integration of DenseNet in U-Net, which involved a multi-path densely connected encoder and inception module-based convolution blocks with dilated convolution at different scales. However, the network input only accepts 2D slides and not 3D volumes.

As reviewed in [3], dense connection-based layer-level fusion improves the effectiveness and efficiency of multi-modal segmentation network through better information propagation, implicit deep supervision and reduced risk of over-fitting on small datasets. While recognising the advantages provided by densely connected networks for multimodal fusion, HyperDense-Net architecture has some limitations which we address in this paper. The first lies in the variation of filter depth. Compared to many other segmentation networks such as U-Net, HyperDense-Net contains no pooling layer between convolutional layers and is overall not so deep (i.e. contains nine convolution blocks and four fully-convolutional layers). However, it retained the conventional way of increasing the number of filters (just like the networks with pooling layers) by doubling or multiplying 1.5 after every three consecutive convolution blocks, resulting in a drastic change in feature abstraction in the 4th and 7th layers and moderate learning in other layers. The other lacking aspect we identified is the way multi-modal feature maps concatenate. In HyperDense-Net, the feature maps from all modalities as well as previous layers are simply fused using concatenation along the channel dimension. We speculate this approach fails to consider the discrepancy in visual features under different modalities and the importance of modal-specific learning, resulting in ineffective multi-modal feature merging and propagation.

Given the challenges and limitations described above, we propose a new densely connected fusion architecture, which we refer to as HyperFusionNet, for multi-modal brain tissue segmentation. The proposed network is trained in an end-to-end fashion, where a progressive feature abstraction process is ensured, and a better feature fusion strategy is integrated to alleviate the interference and incompatibility of feature maps generated from different modality paths. We compare the proposed architecture to the state-of-art method using both a private dataset on acute ischemic stroke lesions and data from the iSeg-2017 MICCAI Grand challenge [18] on 6-month infant brain MRI Segmentation.

2   Method

2.1   Baseline Architecture

The pipeline of the baseline architecture – HyperDense-Net [13] is shown in Fig. 1, but without the added ResFuse modules. Taking the fusion of three modalities as an example, each imaging modality has its own stream for the propagation of the features until it reaches the fully convolutional layer. Every convolutional block includes batch normalization, PReLU activation and convolution with no spatial pooling. For a convolutional block in a conventional CNN, the output of the current layer, denoted as x_l, is obtained by applying a mapping function $F_l(\cdot)$ to the output x_{l-1} of the previous layer, i.e.

$$x_l = F_l(x_{l-1}) \tag{1}$$

However, in the HyperDense-Net, feature maps generated from different modalities as well as the feature outputs from previous layers are concatenated in a feed-forward manner to be input to the convolution block. Let M represents the number of modalities involved in the multi-modal network, the output of the lth layer along a stream $m = 1, 2, \ldots, M$ in the baseline architecture is then defined as.

$$x_l^m = F_l([x_{l-1}^1, x_{l-1}^2, \ldots x_{l-1}^M, x_{l-2}^1, x_{l-2}^2, \ldots, x_{l-2}^M, \ldots, x_0^M]) \tag{2}$$

All streams are then concatenated together before entering the fully convolutional layers. The output of the network is fed into a softmax function to generate the probabilistic map. The final segmentation result is computed based on the highest probability value. The baseline network is optimised using Adam optimiser and cross-entropy loss function.

2.2   Proposed Architecture

To avoid drastic changes of feature abstraction, we first modified the number of filter sizes in the baseline network. Instead of having equal number of filters for every three consecutive convolutional blocks, we gradually increase the number of filters in the successive blocks. Let w denotes the increased value in filter number in the original network, we add $w/3$ filters to the successive convolutional layers in the proposed network. The effectiveness of such design is demonstrated previously in [19].

To improve the fusion of the multi-modal features along each modality path, we propose to merge the feature maps via a 'ResFuse Module'. Inspired by [20], the module (illustrated in Fig. 2) contains a residual connection where the main information belonging to that specific modality path is traversed directly. A 1×1 convolutional layer is also introduced to allow some comprehension of channel correspondence between the features of the specific path and the merging information from other modalities. For the concatenated feature maps, we apply non-linear PReLU activation before summation in order to promote better mapping and information flow in the fused propagation. Equation 2 is then updated to

$$x_l^m = H_l\left(x_{l-1}^m\right)$$
$$+ G_l\left(\left[x_{l-1}^1, x_{l-1}^2, x_{l-1}^M, x_{l-2}^1, x_{l-2}^2, \dots, x_{l-2}^M, \dots, x_0^M\right]\right) \tag{3}$$

where H_l applies the dimension expansion of x_{l-1}^m via the 1×1 convolution and G_l performs the concatenation of features from all modalities and the activation.

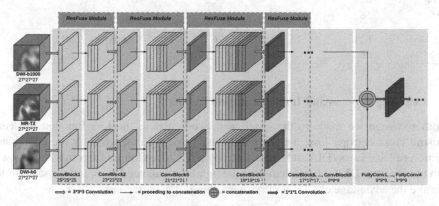

Fig. 1. The proposed HyperFusionNet architecture in the case of three imaging modalities. The feature map generated by each convolutional block is colour coded; the deeper the colour the deeper the layer. The stacked feature maps show how the dense connection and layer shuffling happens originally along each path. The ResFuse Module is added to replace the original concatenation.

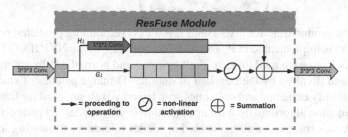

Fig. 2. Proposed residual fusion module for the multi-modal feature merging.

The layer details are presented in Table 1, which shows the layer parameters involved in the proposed network. The overall architecture layout is presented in Fig. 1, which we term HyperFusionNet.

Table 1. The HyperFusionNet architecture detail. Notations: CB - convolutional block; RFM – residual fusion module; FC - fully convolutional layer.

Network components	No. filters	Output size	Network components	No. filters	Output size
CB1	25	25^3	RM6 CB7	819 75	13^3
RFM1 CB2	75 33	23^3	RM7 CB8	1044 83	11^3
RFM2 CB3	174 41	21^3	RM8 CB9	1293 91	9^3
RM3 CB4	297 50	19^3	RM9 FC1	1566 600	9^3
RM4 CB5	447 58	17^3	FC2	300	9^3
			FC3	150	9^3
RM5 CB6	621 66	15^3	FC4	No. classes	9^3

2.3 Learning Process and Implementation Details

Another change we made to the baseline network was to the loss function. Instead of using cross entropy loss, we propose to use Combo Loss, which is the combined function of Dice Loss (DL) and Cross-Entropy (CE). The Combo Loss function allows us to benefit from DL for better handling the lightly imbalanced class and the same time leverage the advantage of CE for curve smoothing. It is defined as

$$L = \alpha \left(-\frac{1}{N} \sum_{i=1}^{N} \beta(g_i \log s_i) + (1-\beta)\left[(1-g_i)\log(1-s_i)\right] \right)$$
$$- (1-\alpha)\left(\frac{2\sum_{i=1}^{N} s_i g_i + \varepsilon}{\sum_{i=1}^{N} s_i + \sum_{i=1}^{N} g_i + \varepsilon} \right) \tag{4}$$

where g_i is the ground truth for pixel i, and s_i is the corresponding predicted probability.

The model is implemented in PyTorch and trained on a single NVIDIA GTX 1080Ti GPU. Images from each modality are skull stripped and normalized by subtracting the mean value and dividing by the standard deviation. 3D image patches of size $27 \times 27 \times 27$ are randomly extracted and only ones with lesion voxels are used for training. The Adam optimization algorithm used for optimization is set with default parameter values. The network was trained for 600 epochs. For model inference, the testing images are first normalised and non-overlapping 3D patches are extracted. The output, which is the $9 \times 9 \times 9$ voxel-wise classification obtained from the prediction at the centre of the patch, is used to reconstruct the full image volume by reversing the extraction process. The source code for the implemented model is available on GitHub[1].

[1] https://github.com/Norika2020/HyperFusionNet.

3 Experiments and Results

3.1 Datasets

The proposed HyperFusionNet is evaluated both on a hospital-collected multi-modal dataset of acute stroke lesion segmentation and on the public iSeg-17 MICCAI Grand Challenge dataset. The hospital-collected dataset was divided into 90 training cases and 30 testing cases, with three modalities in each case, i.e., T2, DWI-b1000 and DWI-b0. All images are of size $256 \times 256 \times 32$. The ground truth for the acute stroke lesion in the dataset is annotated by experienced physicians and there are two classes involved: lesion and non-lesion. Comparably, iSEG17 is a much smaller dataset containing 10 available volumes with two modalities, i.e., T1- and T2- weighted. To be consistent with the experiment carried out in the original baseline paper [13], we also split the dataset into training, validation and testing sets, each having 6, 1, 3 subjects, respectively. There are four classes involved in iSeg-17 dataset, i.e., background, cerebrospinal fluid (CSF), grey matter (GM) and white matter (WM).

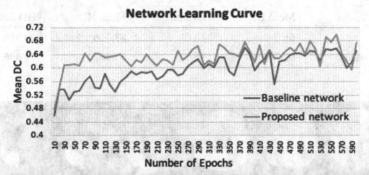

Fig. 3. Validation accuracy measured using mean DC during proposed model training on the stroke lesion dataset.

3.2 Results and Discussion

The proposed network is first evaluated by assessing its performance at segmenting acute stroke lesions in the hospital-collected dataset. In this experiment, the batch size was set to 10 and learning rate was set to 0.0002. Figure 3 shows the comparison of the validation accuracy between the baseline and HyperFusionNet. The mean Dice score of the validation set is calculated after every ten epochs. We can see from the learning curve that HyperFusionNet is not only more accurate compared to the baseline but also converges faster. This can be attributed to the synergy between the residual connections and the feature activation after concatenation. Table 2 shows the segmentation results on the testing volumes in metrices Dice coefficient (DC) and Hausdorff distance (HD). Both measurements suggest that the proposed network provides more effective fusion of multi-modal features than the original approach. Figure 4 shows some examples of qualitative results on three kinds of stroke lesion conditions: a big lesion, multiple lesions and a small lesion. Overall, we observe that the proposed network is better at discarding outliers and predict stroke lesion regions of higher quality.

To better understand how the proposed modifications to the baseline contribute to the network performance, we also did an ablation study. In this experiment, the 3D networks were changed to 2D (i.e. slice-by-slice input with patch size 27×27) to save training and computation time. As shown in Table 3, the accuracy is immediately decreased when the network is changed to 2D. This is expected and it also emphasises the importance of exploiting the slice dimension information for such networks. The results show the clear improvements made by each modification to the 2D baseline network, with ResFuse module making the biggest contribution. We also tested other loss functions – Dice Loss, Focal Loss and Tversky Loss. Comparably, Combo Loss has shown to be more advantageous in our proposed network.

Table 2. The testing results on stroke lesion segmentation measured in DC (%) and HD with their associated standard deviation for the experimented networks.

Network	Mean DC	DC Std	Mean HD	HD Std
Baseline	65.6	18.0	87.756	20.386
HyperFusionNet	67.7	16.5	85.462	14.496

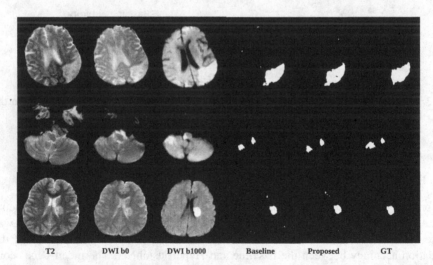

T2   DWI b0   DWI b1000   Baseline   Proposed   GT

Fig. 4. Qualitative results obtained for the stroke dataset using the baseline and the proposed networks.

We also tested the HyperFusionNet on the iSeg-17 dataset to investigate its performance on a smaller dataset with more classes involved. To allow a fair comparison, the parameters such as batch size (=5) and learning rate (initially = 0.001 and reducing by a factor of 2 every 100 epochs) are set to match the baseline paper. The results for the

Table 3. The testing results of the proposed modification to the baseline on stroke lesion segmentation measured in DC (%).

Network modification	DC	Other loss function	DC
Baseline 2D (CE loss)	41.9	HyperFusionNet (CE loss)	46.6
+ Incremental filters	43.0	+ Dice loss	45.9
+ ResFuse module	46.6	+ Focal loss	39.0
+ Combo loss	**47.1**	+ Tversky loss	43.6

Table 4. The performance comparison on the testing set of the iSeg17 brain segmentation measured in DC (%).

Architecture	CSF	WM	GM
Baseline	93.4 ± 2.9	89.6 ± 3.5	87.4 ± 2.7
HyperFusionNet	93.6 ± 2.5	90.2 ± 2.2	87.8 ± 2.3

baseline are reproduced using their published code written in PyTorch[2] in order to compare results under the same experimental setting. Results in Table 4 shows the proposed network yields better segmentation results than the baseline. Although there is not a significant improvement in the averaged Dice score, we observed that it worked well for challenging cases of segmenting GM and WM. Figure 5 depicts such a challenging example where the proposed HyperFusionNet shows a better contour recovery than that obtained by the baseline.

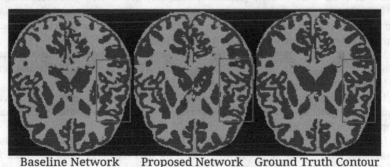

Baseline Network   Proposed Network   Ground Truth Contour

Fig. 5. Qualitative results achieved by the baseline and proposed network compared to the ground truth contour.

[2] https://github.com/josedolz/HyperDenseNet_pytorch.

4 Conclusion

In this work, we propose a novel method HyperFusionNet for brain segmentation using 3D images captured with multiple modalities. The proposed network presents a new way to fuse features from different modalities in a densely connected architecture. A progressive feature abstraction process is promoted and a ResFuse module is introduced to replace the simple concatenated fusion used in the baseline network. The network is improved further with a Combo loss function. We evaluate the proposed network in both ischemic acute lesion segmentation and infant brain segmentation and compare it to a state-of-art multi modal fusion network. The experimental results demonstrate the effectiveness of HyperFusionNet and its capability to tackle challenging multi-modal segmentation tasks with different applications and dataset sizes. Our research largely focused on the fusion network itself, and little data augmentation and post processing was included. For future work, we will improve the network further by implementing pre and post enhancements. The influence of each modality on different applications will also be investigated.

References

1. Tongxue, Z., Su, R., Stéphane, C.: A review: deep learning for medical image segmentation using multi-modality fusion. Array 3–4 (2019)
2. Lapuyade-Lahorgue, J., Xue, J.H., Ruan, S.: Segmenting multi-source images using hidden markov fields with copula-based multivariate statistical distributions. IEEE Trans. Image Process. **26**(7), 3187–3195 (2017)
3. Balasubramaniam, P., Ananthi, N.: Image fusion using intuitionistic fuzzy sets. Inf. Fus. **20**(1), 21–30 (2014)
4. Guo, Z., Li, X., Huang, H., Guo, N., Li, Q.: Deep learning-based image segmentation on multimodal medical imaging. IEEE Trans. Radiat. Plasma Med. Sci. **3**(2), 162–169 (2019)
5. Havaei, M., et al.: Brain tumor segmentation with deep neural networks. Med. Image Anal. **35**, 18–31 (2017)
6. Kamnitsas, K., et al.: Efficient multi-scale 3D CNN with fully connected CRF for accurate brain lesion segmentation. Med. Image Anal. **36**, 61–78 (2017)
7. Lavdas, I., et al.: Fully automatic, multiorgan segmentation in normal whole body magnetic resonance imaging (MRI), using classification forests (CFs), convolutional neural networks (CNNs), and a multi-atlas (MA) approach. Med. Phys. **44**(10), 5210–5220 (2017)
8. Cai, H., Verma, R., Ou, Y., Lee, S., Melhem, E.R., Davatzikos, C.: Probabilistic segmentation of brain tumors based on multi-modality magnetic resonance images. In 4th IEEE International Symposium on Biomedical Imaging, pp. 600–603 (2007)
9. Klein, S., van der Heide, U.A., Lips, I.M., van Vulpen, M., Staring, M., Pluim, J.P.: Automatic segmentation of the prostate in 3D MR images by atlas matching using localized mutual information. Med. Phys. **35**(4), 1407–1417 (2008)
10. Menze, B.H., et al.: The multimodal brain tumor image segmentation benchmark. IEEE Trans. Med. Imaging **34**(10), 1993–2024 (2015)
11. Aygun, M., Sahin, Y.H., Unal, G.: Multimodal convolutional neural networks for brain tumor segmentation. arXiv preprint:1809.06191 (2018)
12. Chen, Y., Chen, J., Wei, D., Li, Y., Zheng, Y.: OctopusNet: a deep learning segmentation network for multi-modal medical images. In: Li, Q., Leahy, R., Dong, B., Li, X. (eds.) MMMI 2019. LNCS, vol. 11977, pp. 17–25. Springer, Cham (2020). https://doi.org/10.1007/978-3-030-37969-8_3

13. Dolz, J., Gopinath, K., Yuan, J., Lombaert, H., Desrosiers, C., Ben Ayed, I.: HyperDense-Net: a hyper-densely connected CNN for multi-modal image segmentation. IEEE Trans. Med. Imaging **38**(5), 1116–1126 (2019)
14. Tseng, K.L., Lin, Y.L., Hsu, W., Huang, C.Y.: Joint sequence learning and cross-modality convolution for 3D biomedical segmentation. In: IEEE Computer Society Conference on Computer Vision and Pattern Recognition (CVPR 2017), pp. 3739–3746 (2017)
15. Dou, Q., Liu, Q., Heng, P.A., Glocker, B.: Unpaired multi-modal segmentation via knowledge distillation. IEEE Trans. Med. Imaging **39**(7), 2415–2425 (2020)
16. Huang, G., Liu, Z., van der Maaten, L., Weinberger, K.Q.: Densely connected convolutional networks. In: IEEE Computer Society Conference on Computer Vision and Pattern Recognition (CVPR 2017), pp. 2261–2269 (2017)
17. Dolz, J., Ben Ayed, I., Desrosiers, C.: Dense multi-path U-net for ischemic stroke lesion segmentation in multiple image modalities. In: Crimi, A., Bakas, S., Kuijf, H., Keyvan, F., Reyes, M., van Walsum, T. (eds.) BrainLes 2018. LNCS, vol. 11383, pp. 271–282. Springer, Cham (2019). https://doi.org/10.1007/978-3-030-11723-8_27
18. Wang, L., et al.: Benchmark on automatic 6-month-old infant brain segmentation algorithms: the iSeg-2017 challenge. IEEE Trans. Med. Imaging **38**(9), 2219–2230 (2019)
19. Drozdzal, M., Vorontsov, E., Chartrand, G., Kadoury, S., Pal, C.: The importance of skip connections in biomedical image segmentation. In: Deep Learning and Data Labeling for Medical Applications, pp. 179–187 (2016)
20. Ibtehaz, N., Sohel Rahman, M.: MultiResUNet: rethinking the U-Net architecture for multimodal biomedical image segmentation. Neural Netw. **121**, 74–87 (2020)

Robust Hydrocephalus Brain Segmentation via Globally and Locally Spatial Guidance

Yuanfang Qiao, Haoyi Tao, Jiayu Huo, Wenjun Shen, Qian Wang, and Lichi Zhang[✉]

School of Biomedical Engineering, Shanghai Jiao Tong University, Shanghai, China
lichizhang@sjtu.edu.cn

Abstract. Segmentation of brain regions for hydrocephalus MR images is pivotally important for quantitatively evaluating patients' abnormalities. However, the brain image data obtained from hydrocephalus patients always have large deformations and lesion occupancies compared to the normal subjects. This leads to the disruption of the brain's anatomical structure and the dramatic changes in the shape and location of the brain regions, which poses a significant challenge to the segmentation task. In this paper, we propose a novel segmentation framework, with two modules to better locate and segment these highly distorted brain regions. First, to provide the global anatomical structure information and the absolute position of target regions for segmentation, we use a dual-path registration network which is incorporated into the framework and trained simultaneously together. Second, we develop a novel Positional Correlation Attention Block (PCAB) to introduce the local prior information about the relative positional correlations between different regions, so that the segmentation network can be guided in locating the target regions. In this way, the segmentation framework can be trained with spatial guidance from both global and local positional priors to ensure the robustness of the segmentation. We evaluated our method on the brain MR data of hydrocephalus patients by segmenting 17 consciousness-related ROIs and demonstrated that the proposed method can achieve high performance on the image data with high variations of deformations. Source code is available at: https://github.com/JoeeYF/TBI-Brain-Region-Segmentation.

Keywords: Image segmentation · Image registration · Hydrocephalus

1 Introduction

Hydrocephalus is an abnormal accumulation of cerebrospinal fluid (CSF) in the patient's brain with persistent ventricular dilatation, which is a secondary injury from Traumatic Brain Injury (TBI). It is usually caused by the obstruction of cerebrospinal fluid pathways, impaired cerebrospinal fluid circulation, etc., which

Electronic supplementary material The online version of this chapter (https://doi.org/10.1007/978-3-030-87586-2_10) contains supplementary material, which is available to authorized users.

A. Abdulkadir et al. (Eds.): MLCN 2021, LNCS 13001, pp. 92–100, 2021.
https://doi.org/10.1007/978-3-030-87586-2_10

leaves the patient in a state of impaired consciousness [3]. The evaluations of the brain's abnormalities and the correspondence to the patient's consciousness state play an important role to assist the clinical assessments of the disease progression. Specifically, it is demonstrated in [8] that there are 17 brain regions whose functional and anatomical shape states have certain correlations with the improvements of the consciousness level. Therefore, the identification and parcellation of these consciousness-related brain regions are demanding. However, it is generally impractical to conduct segmentation manually, which is tedious, time-consuming, and introduces inter-observer variability. The issues become more deteriorated for hydrocephalus brain images, which contain even higher variability and extent brain changes than the normal brain images. Therefore, the development of an accurate and automatic brain parcellation method for hydrocephalus images would be highly beneficial.

With the development of deep learning, the convolutional neural network (CNN) and its extensions have dominated the field of medical image segmentation. Specifically, UNet [5] combined the high-level and low-level features with different context information to estimate precise segmentation results, and has become the most applied method in this field. Many attempts were also made specifically for brain image segmentation. For example, Moeskops et al. [11] used a CNN to achieve the automatic segmentation of MR brain images into some tissue classes. Ghafoorian et al. [6] integrated the anatomical location information into the network to get explicit location features and improve the segmentation results substantially. The alignment-based brain mapping methods, which adopt the registration techniques such as VoxelMorph [2] to align the brain template to the target for brain parcellation, have also been widely applied.

However, most brain region segmentation methods are designed for normal brain images, while the hydrocephalus images have much higher variations of anatomical structures compared to the normal, due to the large deformations and lesion occupancies caused by the diseases as shown in Fig. 1. The brain anatomical structure information is much more complicated to be encoded for constructing the segmentation model, especially when the training data are also limited. Some attempts have been made to resolve these issues in hydrocephalus brain segmentation. Ledig et al. [10] intended to develop a multi-atlas-based method to segment the brain regions on TBI data, but the experiments reported the failed cases. Ren et al. [12] proposed a two-stage framework with hard and soft attention modules to segment the brain regions of hydrocephalus brain images and demonstrated that it outperforms the state-of-art methods such as UNet. However, it is not an end-to-end framework, where the two modules are not fully integrated into a single network to fully share their conducted features.

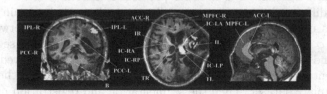

Fig. 1. Exemplar hydrocephalus MR brain images with consciousness-related regions.

In this paper, we intend to parcellate the 17 consciousness-related regions according to [8,12] from the MR hydrocephalus images. To resolve the issues in hydrocephalus brain segmentation, we focus on developing an end-to-end novel framework, with two modules to locate and segment the target regions from two different perspectives: (1) We use a registration guidance module to produce the segmentation network with more anatomical structure information about the absolute position of target regions in the whole brain. (2) We propose a novel Positional Correlation Attention Block (PCAB) integrated into the UNet to improve performance by conducting more explicitly structural features. In the PCAB, we design a Positional Correlation Layer (PCL) to extract the relative positional relationship between different brain regions and use it to refine the segmentation estimation via the attention layer. The absolute and relative position information can provide complementary and comprehensive guidance to the segmentation network, which is designed as an end-to-end network for better information integration. Experiments show that it can outperform the alternatives with a statistical significance, which is evaluated by 5-fold cross-validation on 17 brain regions with great deformations from the collected hydrocephalus brain images.

2   Method

To reduce the impact of the deformation caused by the occupancy and erosion of lesion areas on normal brain areas, we propose two novel modules to extract more comprehensive anatomical structure information from the absolute location in the whole brain and the correlation with each other's location. Figure 2 shows our proposed segmentation framework with dual-path registration module and Positional Correlation Attention Block (PCAB) module. The two networks are trained simultaneously but only the segmentation network is used in the inference stage.

2.1   Guidance with Registration Module

We use an UNet-like network with a dual-path encoder to non-rigidly align the hydrocephalus subject (moving set) with the brain template (fixed set), which is shown in Fig. 2(b). Specifically, we adopt a dual-path encoder that takes the original brain image (I_M, I_F) and the brain segmentation mask (M_M, M_F) as two channels input to make the registration network focus on the ROI regions. The features extracted from the encoder are concatenated with the corresponding decoder features by skip connections to make a low-level and high-level feature fusion. The decoder predicts a deformation field ϕ to align the mask of different brain regions. Finally, the hydrocephalus brain image and brain regions mask are warped using Spatial Transformation Network (STN) according to ϕ. To introduce more supervised information into the segmentation framework and improve the robustness [13], we warp the output prediction map of the segmentation network according to ϕ above and calculate the similarity with the template dataset. Then the segmentation and registration network can be trained simultaneously.

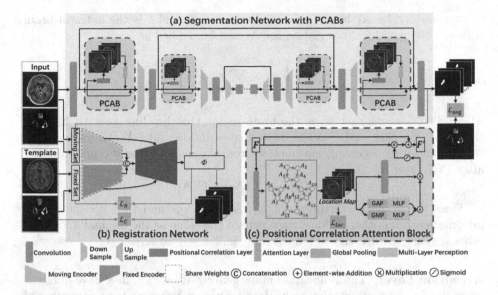

Fig. 2. The overview of the proposed method. The framework consists of a segmentation network and a registration network. The Positional Correlation Attention Blocks (PCAB) are integrated into the segmentation network. Note that only the segmentation network is used in the inference stage.

2.2 Segmentation with Positional Correlation Attention Block

We use a 3D-UNet as our main segmentation network, and PCAB is integrated into different levels. The purpose of PCAB is to use the location correlation of different brain regions to generate the attention map with anatomical structure information. The PCAB includes a Positional Correlation Layer (PCL) to estimate the location map and an attention layer to generate the attention map to refine the input feature. As shown in Fig. 2(a), the PCAB is a plug-and-play block and is integrated into the first two levels of the encoder and the last two levels of the decoder in our study.

Positional Correlation Layer. Localization probability maps for each brain region $A \in \mathbb{R}^{l \times w \times h \times 17}$ are calculated from the input feature $F \in \mathbb{R}^{l \times w \times h \times c}$ by a convolutional layer. Our PCL forms the target brain regions into two directed graph structures according to their correlation with each other's location and the graph is shown in Fig. 2(c). The probability maps for every brain region are passed to the adjacent region using a convolutional layer according to the linkage between each brain region in the graph above. The method has been used in pose estimation [4]. To further expand receptive fields, we use the 3D dilated convolution with dilation rate of 2 and the filter size of 7×7. Note that due to the large distance between some regions, we exclude 3 brain regions and constructed 2 directed graph structures using the remaining 14 brain regions.

Specifically, let A_k be the original feature maps and k is the index of brain regions. The positional correlation for A_k can be defined as

$$\tilde{A}_k = f\left(A_k + \sum_{j\in N} C(A_j)\right), \tag{1}$$

where $j \in N$ means that A_j is the brain region that has linkage with A_k, f is the ReLU function, and C is the dilated convolution. Take A_{11} as an example, A_{11} is refined by receiving information from $\tilde{A}_9$ and A_{10}, so the updated $\tilde{A}_{11}$ after PCL layer is

$$\tilde{A}_{11} = f(A_{11} + C(\tilde{A}_9) + C(A_{10})). \tag{2}$$

Since the graph is starting from A_1, A_5 and A_{10}, and they do not receive information from other regions, they remain the same as the original ones. The other region's feature map is refined in a similar way to A_{11}.

Attention Layer. The 17 location maps generated by PCL may have different priorities for the segmentation task. Here we use an attention layer consisting of spatial attention and channel attention to exploiting the most significant features to refine the input feature. First, we use a convolution layer with an output channel number of c to generate a feature map $\alpha \in \mathbb{R}^{l\times w\times h\times c}$, which has the same size as the input feature. Next, we use a global max pooling (GMP) and a global average pooling (GAP) to get the global information of each channel, which are represented as $F_{max} \in \mathbb{R}^{1\times1\times1\times17}$ and $F_{avg} \in \mathbb{R}^{1\times1\times1\times17}$, respectively. Then, the channel attention coefficient $\beta \in \mathbb{R}^{1\times1\times1\times c}$ is calculated by a multiple layer perception consisting of two fully connected layers and a ReLU activation. The output feature of PCAB $\tilde{F}$ can be obtained as $\tilde{F} = F + \sigma(\alpha \cdot \beta) \cdot F$ and is fed into the next convolution block, where σ denotes a sigmoid function.

2.3 Training Strategy

The registration network loss consists of three following components:

$$\mathcal{L}_R = \beta_1 \mathcal{L}_{NCC}(I_M(\phi), I_F) + \beta_2 \mathcal{L}_{mask}(M_M(\phi), M_F) + \beta_3 \mathcal{L}_{smooth}(\phi), \tag{3}$$

where the first term is local cross-correlation loss between the warped image and fixed image which is the same as VoxelMorph, the second term calculates the mean square error (MSE) between the brain region mask, and the last term penalizes local spatial variations ϕ to keep it smooth.

The segmentation network is optimized by two aspects of supervision information. The loss function is written as follows:

$$\mathcal{L}_S = \alpha_1 \mathcal{L}_{seg}(\hat{Y}, Y_G) + \alpha_2 \frac{1}{N} \sum_{i=1}^{N} \mathcal{L}_{loc}^i(\tilde{A}^i, A_G^i). \tag{4}$$

The first term in Eq. (4) calculates the cross-entropy between the prediction map of the segmentation network $\hat{Y}$ and the ground-truth Y_G. The second term calculates the weighted binary-cross-entropy between the location maps $\tilde{A}^i$ from PCL at each level i and the ground-truth A_G^i. A_G^i is the center of each brain region which is dilated into ball areas with a radius of 3. To balance the loss of foreground and background, we use the inverse of the average pixel value of ground-truth as the loss weight of the foreground. The equation for the second term is shown below, where V^i denotes the sum of voxels for level i:

$$\mathcal{L}_{\text{loc}}^i(\tilde{A}, A_G) = -\frac{V^i}{\sum A_G^i} \cdot A_G^i \cdot \log \tilde{A}^i - \left(1 - A_G^i\right) \cdot \log\left(1 - \tilde{A}^i\right). \tag{5}$$

During the training stage, we warp the segmentation map according to ϕ and calculate the MSE with brain template ROIs, which is defined as

$$\mathcal{L}_C = \gamma \mathcal{L}_{\text{mask}}(\hat{Y}(\phi), M_F), \tag{6}$$

where we use $\mathcal{L}_C$ to train the segmentation network and registration network simultaneously, but only when the training epoch is greater than 15 to keep the training process stable. In this way, the overall loss function can be defined as $\mathcal{L} = \mathcal{L}_S + \mathcal{L}_R + \mathcal{L}_C$. Note that β_1, β_2, β_3 in Eq. (3), α_1, α_2 in Eq. (4) and γ in Eq. (6) are hyper-parameters and γ is 0 when the train epochs are less than 15.

3 Experiments and Results

3.1 Datasets and Experiments

The proposed method was evaluated on in-house MR brain data in T1 from 44 hydrocephalus patients and all subjects have hematoma volume and hydrocephalus disease. The brain template used in the registration network is Colin 27 Average Brain [7]. The 17 consciousness-related ROIs shown in Fig. 1 was manually delineated and used for training. Before the data was fed into the network, we preprocessed the data by the following steps: The images' voxel spacing was resized to $1\,\text{mm} \times 1\,\text{mm} \times 1\,\text{mm}$. Then the histogram matching was conducted for intensity normalization. In the registration module, the moving set was firstly affine-registered to the fixed set using the ANTs package [1]. The data were grouped into 5-fold for the cross-validation and ablation study.

The network was implemented using PyTorch1.6 and trained on NVIDIA RTX Titan GPU. We used different learning rate settings for the two networks: 1e−3 for the registration network and 1e−4 for the segmentation network. All learning rate settings have a decay of 0.1 every 5 epochs and the epochs of training are 100. Adam optimizer was adopted with a weight decay of 1e−4. Due to the limitation of the GPU's memory size, the batch size was set to 1. Therefore, we replaced the BatchNorm with GroupNorm to reduce the effect of small batch size, and the number of groups were set to 4.

3.2 Results

To verify the effect of the proposed method, we make the ablation studies by adjusting the number of PCABs denoted as N in Eq. (4) and the incorporation of

Table 1. Dice Coefficient of ablation studies and comparison with other methods. PCABs(N) means that there are N PCABs integrated into the framework.

Methods	Dice coefficient(%)
(a) Ablation studies	
UNet(Baseline)	61.64 ± 21.60
UNet+Registration	63.85 ± 18.71
UNet+PCABs(2)	64.83 ± 18.67
UNet+PCABs(4)	66.00 ± 17.76
UNet+PCABs(4)+Registration	**69.03 ± 14.85**
(b) Comparison with other methods	
VoxelMorph[2]	39.23 ± 22.23
UNet [5]	61.64 ± 21.60
nnUNet [9]	63.37 ± 23.62
Ren et al. [12]	67.19 ± 17.18
Proposed	**69.03 ± 14.85**

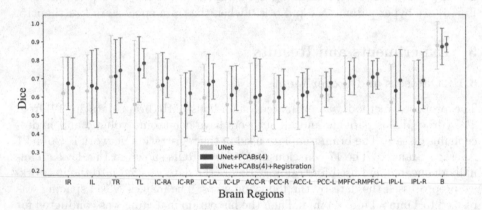

Fig. 3. Dice Coefficient of each brain region in different studies. (Color figure online)

the registration network, which are shown in Table 1(a). PCABs(2) means that the PCAB is only integrated into the first level and the last level of decoder, while PCABs(4) integrated two more PCABs in the deeper levels like Fig. 2. We use Dice Coefficient Score (DCS) as the evaluation metrics.

As presented in Table 1(a), the DSC score is increased by 4.36% by applying PCAB into baseline UNet, and with the growth of the number of PCABs the performance also increases. The results indicate that the PCABs can help the network to encode the variations of the anatomical structure by extracting the positional correlations information. By training the segmentation network and registration network together, the DSC further improved 3.03% and the standard deviation decreased 2.91% which proved that the registration network further confirmed the absolute position of the regions in the whole brain. The results also show that the both two networks helps to achieve the best segmentation

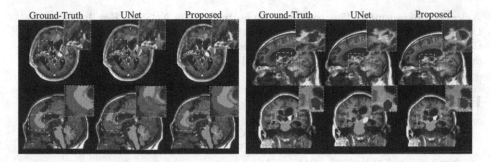

Fig. 4. Visualization of segmentation results comparison of UNet and the proposed method.

performance. Figure 3 shows the distribution of dice scores of each brain region in the different ablation studies. It can be concluded that the method proposed has better performance than the baseline method. The final proposed methods (red) have higher average dice and lower standard deviation in most of the brain regions, indicating that our method can achieve more accurate and more stable results for the hydrocephalus dataset.

We also compared our method with other state-of-the-art segmentation methods. The results are shown in Table 1(b). Our method achieved the average dice of 69.03% with a standard deviation of 14.85%, which is 1.84% higher than the recent state-of-the-art method [12]. The lower standard deviation also indicates that the result has improved more on the hard subject samples which have larger deformation than the others, where the proposed method achieves more robust performance than the alternatives.

Figure 4 visualizes the results from our segmentation framework and the UNet method. The first column is ground-truth and the next column are the results of UNet and the method proposed, respectively. As shown in the zoomed block, our method can locate and segment the brain regions, while the UNet fails to handle. For the cases with large deformation, there is a huge improvement in accuracy and robustness with our proposed method.

4 Conclusion

We have proposed a novel segmentation framework based on globally registration guidance and locally relative positional correlations guidance for the hydrocephalus dataset. First, we used a dual-path registration network to provide global anatomical structure information which was important to the final segmentation results. Besides, to make the network focus more on the target regions, we integrated a Positional Correlation Attention Block (PCAB) into UNet to generate a location attention map to refine the feature of the encoder and decoder. The comparative results show that the proposed method can achieve higher results and have a greater improvement for cases with larger deformation, indicating that our method has high accuracy and robustness.

Acknowledgements. This work was supported by the National Key Research and Development Program of China (2018YFC0116400), National Natural Science Foundation of China (NSFC) grants (62001292), Shanghai Pujiang Program (19PJ1406800), and Interdisciplinary Program of Shanghai Jiao Tong University.

References

1. Avants, B.B., Tustison, N.J., Song, G., Cook, P.A., Klein, A., Gee, J.C.: A reproducible evaluation of ants similarity metric performance in brain image registration. Neuroimage **54**(3), 2033–2044 (2011)
2. Balakrishnan, G., Zhao, A., Sabuncu, M.R., Guttag, J., Dalca, A.V.: VoxelMorph: a learning framework for deformable medical image registration. IEEE Trans. Med. Imaging **38**(8), 1788–1800 (2019)
3. Chari, A., Czosnyka, M., Richards, H.K., Pickard, J.D., Czosnyka, Z.H.: Hydrocephalus shunt technology: 20 years of experience from the Cambridge shunt evaluation laboratory. J. Neurosurg. **120**(3), 697–707 (2014)
4. Chu, X., Ouyang, W., Li, H., Wang, X.: Structured feature learning for pose estimation. In: Proceedings of the IEEE Conference on Computer Vision and Pattern Recognition, pp. 4715–4723 (2016)
5. Çiçek, Ö., Abdulkadir, A., Lienkamp, S.S., Brox, T., Ronneberger, O.: 3D U-net: learning dense volumetric segmentation from sparse annotation. In: Ourselin, S., Joskowicz, L., Sabuncu, M.R., Unal, G., Wells, W. (eds.) MICCAI 2016. LNCS, vol. 9901, pp. 424–432. Springer, Cham (2016). https://doi.org/10.1007/978-3-319-46723-8_49
6. Ghafoorian, M., et al.: Location sensitive deep convolutional neural networks for segmentation of white matter hyperintensities. Sci. Rep. **7**(1), 1–12 (2017)
7. Holmes, C.J., Hoge, R., Collins, L., Woods, R., Evans, A.C.: Enhancement of MR images using registration for signal averaging. J. Comput. Assist. Tomogr. **3**(2), 324–333 (1998)
8. Huo, J., et al.: Neuroimage-based consciousness evaluation of patients with secondary doubtful hydrocephalus before and after lumbar drainage. Neurosci. Bull. (9) (2020)
9. Isensee, F., Jaeger, P.F., Kohl, S.A.A., Petersen, J., Maier-Hein, K.H.: nnU-net: a self-configuring method for deep learning-based biomedical image segmentation. Nat. Methods **18**(2), 203–211 (2021)
10. Ledig, C., et al.: Robust whole-brain segmentation: application to traumatic brain injury. Med. Image Anal. **21**(1), 40–58 (2015)
11. Moeskops, P., Viergever, M.A., Mendrik, A.M., De Vries, L.S., Benders, M.J., Išgum, I.: Automatic segmentation of MR brain images with a convolutional neural network. IEEE Trans. Med. Imaging **35**(5), 1252–1261 (2016)
12. Ren, X., Huo, J., Xuan, K., Wei, D., Zhang, L., Wang, Q.: Robust brain magnetic resonance image segmentation for hydrocephalus patients: Hard and soft attention. In: 2020 IEEE 17th International Symposium on Biomedical Imaging (ISBI), pp. 385–389. IEEE (2020)
13. Xu, Z., Niethammer, M.: DeepAtlas: joint semi-supervised learning of image registration and segmentation. In: Shen, D., Liu, T., Peters, T.M., Staib, L.H., Essert, C., Zhou, S., Yap, P.-T., Khan, A. (eds.) MICCAI 2019. LNCS, vol. 11765, pp. 420–429. Springer, Cham (2019). https://doi.org/10.1007/978-3-030-32245-8_47

Brain Networks and Time Series

Geometric Deep Learning of the Human Connectome Project Multimodal Cortical Parcellation

Logan Z. J. Williams[1,2(✉)], Abdulah Fawaz[2], Matthew F. Glasser[3],
A. David Edwards[1,4,5], and Emma C. Robinson[1,2]

[1] Centre for the Developing Brain, Department of Perinatal Imaging and Health,
School of Biomedical Engineering and Imaging Sciences, King's College London,
London SE1 7EH, UK
logan.williams@kcl.ac.uk

[2] Department of Biomedical Engineering, School of Biomedical Engineering
and Imaging Science, King's College London, London SE1 7EH, UK

[3] Departments of Radiology and Neuroscience, Washington University Medical
School, Saint Louis, MO 63110, USA

[4] Department for Forensic and Neurodevelopmental Sciences, Institute of Psychiatry,
Psychology and Neuroscience, King's College London, London SE5 8AF, UK

[5] MRC Centre for Neurodevelopmental Disorders, King's College London,
London SE1 1UL, UK

Abstract. Understanding the topographic heterogeneity of cortical organisation is an essential step towards precision modelling of neuropsychiatric disorders. While many cortical parcellation schemes have been proposed, few attempt to model inter-subject variability. For those that do, most have been proposed for high-resolution research quality data, without exploration of how well they generalise to clinical quality scans. In this paper, we benchmark and ensemble four different geometric deep learning models on the task of learning the Human Connectome Project (HCP) multimodal cortical parcellation. We employ Monte Carlo dropout to investigate model uncertainty with a view to propagate these labels to new datasets. Models achieved an overall Dice overlap ratio of $>0.85 \pm 0.02$. Regions with the highest mean and lowest variance included V1 and areas within the parietal lobe, and regions with the lowest mean and highest variance included areas within the medial frontal lobe, lateral occipital pole and insula. Qualitatively, our results suggest that more work is needed before geometric deep learning methods are capable of fully capturing atypical cortical topographies such as those seen in area 55b. However, information about topographic variability between participants was encoded in vertex-wise uncertainty maps, suggesting a potential avenue for projection of this multimodal parcellation to new datasets with limited functional MRI, such as the UK Biobank.

Keywords: Human connectome project · Geometric deep learning ·
Cortical parcellation

© Springer Nature Switzerland AG 2021
A. Abdulkadir et al. (Eds.): MLCN 2021, LNCS 13001, pp. 103–112, 2021.
https://doi.org/10.1007/978-3-030-87586-2_11

1   Introduction

Cortical parcellation is the process of segmenting the cerebral cortex into functionally specialised regions. Most often, these are defined using sulcal morphology [5], and are propagated to individuals from a population-average template (or set of templates) based on the correspondence of cortical shape [17,28]. By contrast, while it is possible to capture subject-specific cortical topography from functional imaging in a data-driven way [14], it is difficult to perform population-based comparisons with these approaches as they typically result in parcellations where the number and topography of the parcels vary significantly across subjects [15]. Notably, even following image registration methods that use both structural and functional information [25,26], considerable topographic variation remains across individuals [10,19].

Recently, [10] achieved state-of-the-art cortical parcellation through hand annotation of a group-average multimodal magnetic resonance imaging (MRI) atlas from the Human Connectome Project (HCP). Specifically, a sharp group average of cortical folding, cortical thickness, cortical myelination, task and resting state functional MRI (fMRI), were generated through novel multi-modal image registration [25] driven by 'areal features': specifically T1w/T2w ratio (cortical myelin) [13] and cortical fMRI; modalities which are known to more closely reflect the functional organisation of the brain. This improved alignment allowed for manual annotation of regional boundaries via identification of sharp image gradients, consistent across modalities. With this group average template, they trained a multi-layer perceptron (MLP) classifier to recognise the multimodal 'fingerprint' of each cortical area. This approach allowed [10] to propagate parcellations from labelled to unlabelled subjects, in a registration-independent manner, also providing an objective method to validate parcellation in an independent set of test participants. This classifier detected 96.6% of the cortical areas in test participants, and could correctly parcellate areas in individuals with atypical topography [10].

However, even in this state-of-the-art approach, the classifier was still unable to detect 3.4% of areas across all subjects [10]. Moreover, they were unable to replicate previously identified parcels in regions such as the orbitofrontal cortex [23] and the association visual cortex [1]. It is also unknown whether this classifier generalises to different populations with lower quality data, for example the UK Biobank [21] and the Developing Human Connectome Project [20]. Thus, development of new tools that improve upon areal detection and allow generalisation of this parcellation to new populations with less functional MRI data is warranted. To this end, we consider convolutional neural networks (CNNs), which have proven state-of-the-art for many 2D and 3D medical imaging tasks [3,16]. More specifically, we benchmark a range of different geometric deep learning (gDL) frameworks, since these adapt CNNs to irregular domains such as surfaces, meshes and graphs [2].

The specific contributions of this paper are as follows:

1. We propose a novel framework for propagating the HCP cortical parcellation [10] to new surfaces using gDL methods. These offer a way to improve

over vertex-wise classifiers (as used by [10]) by additionally learning the spatial context surrounding different image features.

2. Since gDL remains an active area of research, with several *complementary* approaches for implementing surface convolutions, we explore the potential to improve performance by ensembling predictions made across a range of models.

3. Given the degree of heterogeneity and anticipated problems in generalising to new data, we return estimates of model uncertainty using techniques for Bayesian deep learning implemented using Monte Carlo dropout [8].

2   Methods

2.1   Participants and Image Acquisition

A total of 390 participants from the HCP were included in this study. Acquisition and minimal preprocessing pipelines are described in [12]. Briefly, modalities included T1w and T2w structural images, task-based and resting state-based fMRI images, acquired at high spatial and temporal resolution on a customized Siemens 3 T (3T) scanner [12]. From these, a set of 110 features were derived and used as inputs for cortical parcellation: 1 thickness map corrected for curvature, 1 T1w/T2w map [13], 1 surface curvature map, 1 mean task-fMRI activation map, 20 task-fMRI component contrast maps, 77 surface resting state fMRI maps (from a $d = 137$ independent component analysis), and 9 visuotopic features. This differs from 112 features used by the MLP classifier in [10] in that artefact features were not included and visuotopic spatial regressors were included. Individual subject parcellations *predicted* by the MLP classifier were used as labels for training each gDL model, as there are no ground truth labels available for multimodal parcellation in the HCP.

2.2   Modelling the Cortex as an Icosphere

For all experiments, the cortical surface was modelled as a regularly tessellated icosphere: a choice which reflects strong evidence that, for many parts of the cortex, cortical shape is a poor correlate of cortical functional organisation [7,10]. Icospheres also offer many advantages for deep learning. Since their vertices form regularly spaced hexagons, icospheric meshes allow consistently shaped spatial filters to be defined and lend themselves to straightforward upsampling and downsampling. This generates a hierarchy of regularly tessellated spheres over multiple resolutions, which is particularly useful as it allows deep learning models to aggregate information through pooling.

2.3   Image Processing and Augmentation

Spherical meshes and cortical metric data (features and labels) for each subject were resampled from the 32k (FS_LR) HCP template space [30], to a sixth-order icosphere (with 40,962 vertices). Input spheres were augmented using non-linear spherical warps estimated by: first, randomly displacing the vertices of a 2nd order icospheric mesh; then propagating these deformations to the input meshes using barycentric interpolation. In total, 100 warps were simulated, and these were randomly sampled from during training. Cortical metric data were then normalised to a mean and standard deviation of 0 and 1 respectively, using precomputed group means and standard deviations per feature.

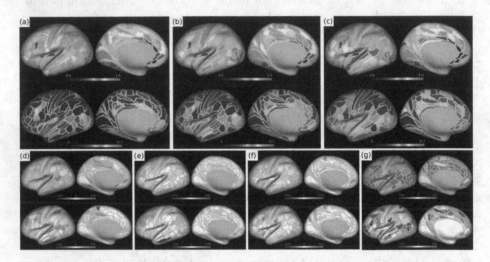

Fig. 1. Mean (top row) and standard deviation (bottom row) (a) Dice overlap ratio, (b) recall score (c) and precision score per region for gDL ensemble. Mean (top row) and standard deviation (bottom row) Dice overlap ratio per region for (d) ensemble - ChebNet, (e) ensemble - GConvNet, (f) GConvNet - MoUNet, and (g) ensemble - Spherical UNet

2.4   Model Architecture and Implementation

Geometric convolutions may be broadly classified into spatial or spectral methods, which reference the domain that the convolution is computed in (see [2,9] for more details). In brief, spatial methods [22,32] simulate the familiar concept of passing a localised filter over the surface. In practice, while expressive, such methods often approximate mathematically correct convolutions; since, due to lack of a single, fixed coordinate system it is not possible to slide a filter over a curved surface whilst maintaining consistent filter orientation. Spectral methods, on the other hand, utilise an alternate representation in which the (generalised) Fourier transform of a convolution of two functions may be represented by the product of their Fourier transforms. As full spectral methods are computationally expensive, it is standard practice to address this through polynomial approximation [4].

Each method therefore results in different compromises, and for that reason offers complementary solutions, which in principle may be combined to improve performance. In this paper, we therefore benchmark and ensemble two spatial networks: *Spherical U-Net* [32] and *MoUNet* [22]; and two spectral (polynomial approximation) methods: *ChebNet* [4] and *GConvNet* [18].

In each case, methods were implemented with a U-Net [27] like architecture with a 6-layer encoder and decoder, and upsampling was performed using transpose convolution (as implemented by [32]). Code for Spherical U-Net was implemented from its GitHub repository[1] and ChebNet, GConvNet and MoUnet were written using PyTorch Geometric [6]. Optimisation was performed using Adam, with an unweighted Dice loss and learning rates: 1×10^{-3} (for Spherical U-Net) and 1×10^{-4} (for all other models). All models were implemented on a Titan RTX 24 GB GPU, with batch size limited to 1 due to memory constraints (resulting from the high dimension of input channels). Models were trained and tested with a train/validation/test split of 338/26/26, using data from both left and right hemispheres. Following training, an unweighted ensemble approach was taken, where one-hot encoded predictions for a single test subject were averaged across all gDL models. Model performance was also assessed using weighted recall and precision scores. Finally uncertainty estimation was implemented using test-time dropout [8] (with $p = 0.2$, the probability of an input channel being dropped). Vertex-wise uncertainty maps were produced by repeating dropout 200 times, and then calculating the standard deviation across each vertex of the predicted parcellation per subject.

3   Results

Table 1 shows the overall performance each model on HCP parcellation using single subject cortical maps predicted by the HCP MLP classifier. All methods perform well, achieving a Dice overlap ratio of >0.85, recall score of >0.82 and precision score of >0.85. The mean and standard deviation Dice overlap ratio, recall and precision scores per area are shown for GConvNet (the best performing model) in Fig. 1a–c. V1 and cortical areas in the parietal lobe had higher a mean and lower standard deviation Dice overlap ratio, whilst cortical areas in the medial frontal lobe, occipital pole and insula had lower a mean and higher standard deviation Dice overlap ratio. At the level of a single cortical region, mean and standard deviation Dice overlap ratio varied across models (Fig. 1b–d), and this regional variability was utilised through an ensemble approach to improve parcellation performance (Table 1).

The ability of gDL models to detect atypical cortical topography was assessed qualitatively in a test-set participant where area 55b was split into three distinct parcels by the frontal and posterior eye fields [11]. This showed that, while none of the gDL models predicted this split (Fig. 2b), vertex-wise uncertainty maps highlighted the split as a region of uncertainty. Figure 3b demonstrates that the most likely labels for this subject (at the vertex marked with a white dot) were the frontal eye fields (184/200 epochs) and area 55b (16/200 epochs). By contrast, when compared to a similar vertex location in a subject with typical

---

[1] https://github.com/zhaofenqiang/Spherical_U-Net.

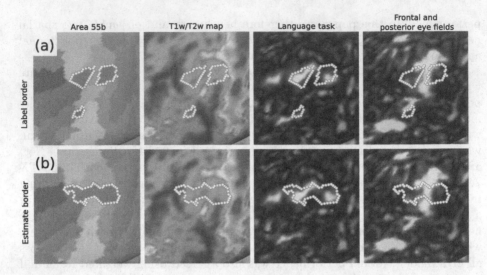

Fig. 2. Label border (a) and estimated (b) border predicted by gDL ensemble for test set participant with atypical area 55b topography. Borders are overlaid on T1w/T2w map, and functional connectivity map from the HCP language task, and functional connectivity map highlighting the frontal and posterior eye fields.

Table 1. Mean ± standard deviation Dice overlap ratio, recall, and precision for all four geometric deep learning methods and the unweighted ensemble approach

Method	Dice overlap ratio	Recall	Precision
ChebNet	0.871 ± 0.021	0.839 ± 0.024	0.862 ± 0.016
GConvNet	**0.875 ± 0.020**	**0.843 ± 0.230**	**0.865 ± 0.015**
MoUNet	0.873 ± 0.021	0.841 ± 0.023	0.864 ± 0.015
Spherical UNet	0.860 ± 0.021	0.825 ± 0.022	0.851 ± 0.013
Ensemble	**0.880 ± 0.019**	**0.848 ± 0.022**	**0.860 ± 0.019**

parcellation in area 55b (Fig. 3c), there was no uncertainty in the estimated label, predicting area 55b across all epochs.

Beyond area 55b, gDL models often predicted cortical areas as single contiguous parcels, whereas the HCP MLP classifier predicted some cortical areas as being comprised of several smaller, topographically-distinct parcels. This uncertainty relative to the MLP is further emphasised by the findings from the Monte Carlo dropout uncertainty modelling which showed that areas of uncertainty tended to be greatest along the boundaries between regions, and were higher in locations where >2 regions met.

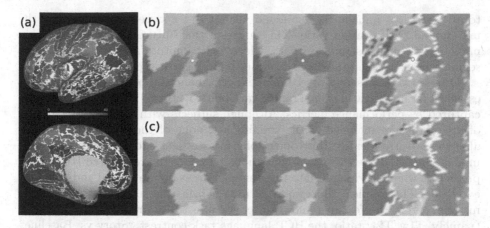

Fig. 3. (a) Example of a vertex-wise uncertainty map produced using Monte Carlo dropout (MoUNet) (b) From left to right: label, estimate and vertex-wise uncertainty map in subject with *atypical* topography of area 55b. (c) from left to right: label, estimate and vertex-wise uncertainty map in subject with *typical* topography of area 55b.

4   Discussion

Developing methods that capture the topographic variability of cortical organisation is essential for precision modelling of neuropsychiatric disorders. Here we show that gDL methods achieve good performance in predicting subject's cortical organisation, when trained on labels output from the HCP MLP classifier.

Even though overall metrics of regional overlap were high, there was marked variability across cortical areas. These findings are in part a consequence using an unweighted Dice loss, since, in this case, mislabelling single vertices of smaller cortical areas will have less impact than for larger ones [24]. This is reflected in the results above, where larger regions e.g. V1, and cortical areas in the parietal lobe, had higher mean and lower standard deviation Dice overlap ratio, recall and precision score per region. This is compared to smaller regions in the medial frontal lobe, insula, and lateral occipital lobe, which had lower mean and higher standard deviation. This inherent limitation of the Dice overlap ratio might also explain why GConvNet (the gDL method with the smallest kernel size) performed the best, as it was capable of learning very localised features. The variation in performance across cortical areas also differed between models, which suggests that each gDL model is learning a different set of features. This was expected given the theoretical differences in how in each model's convolution is defined. Utilising these differences in an ensemble approach improved Dice overlap ratio by 0.005 (0.5%) above GConvNet, which translates to an overlap improvement of 200 vertices on a 6th-order icosphere (on the same icosphere, area 55b is only 123 vertices in size).

Although not described here, we also trained these gDL methods using a generalised (weighted) Dice overlap ratio as described by [29] that is designed to address class imbalance, but found that it did not perform as well as the unweighted Dice overlap ratio. This suggests that future work on improving model performance should, in part, address the limitations of common image segmentation losses in the context of multimodal cortical parcellation.

Qualitative assessment of gDL model performance on cortical parcellation is essential for investigating topographic variability, as this information is not fully captured by performance metrics. Although subjects with atypical topography of area 55b were included in the training set, none of the gDL methods were able to correctly identify this topography in a test-set subject. Specifically, all models predicted area 55b as a contiguous parcel compared to the HCP MLP prediction where it was split into three smaller areas by the frontal and posterior eye fields. The atypical topography of area 55b in this subject was confirmed manually from the features known to contribute to its multimodal fingerprint (namely, T1w/T2w ratio, the HCP language task contrast "Story vs. Baseline" and resting-state functional connectivity map) [11].

The performance of the gDL models in area 55b highlights the overall tendency of these models to predict cortical areas as contiguous regions compared to those predicted by the HCP MLP classifier. This behaviour might be a result of CNNs learning spatial context, and a strong bias towards learning typical topographic organisation due to downsampling and skip connections in the U-Net architecture. In contrast, the HCP MLP was trained to classify each vertex independently using limited spatial context (30 mm radius searchlight across the surface) [10]. The importance of spatial contiguity in defining cortical areas is unknown, but given the lack of ground truth it is difficult to evaluate which approach is more accurate without extensive further qualitative and quantitative evaluation. However, these results do suggest each model introduces unique biases that need to be accounted for when investigating cortical organisation and neuropsychiatric disorders.

Achieving multimodal cortical parcellation in datasets beyond the HCP will be invaluable for precision modelling of neuropsychiatric disorders. However, generalising these multimodal labels to other datasets such as the UK Biobank (healthy ageing adults) [21] and the Developing Human Connectome Project (term and preterm neonates) [20] is challenging due to differences in population demographics and data acquisition (less and lower quality). The vertex-wise uncertainty maps introduced here provide a quantitative method to evaluate label propagation, which also could be used to inform post-processing of individual participant cortical parcellations, similar in nature to [10]. The information about topographic variability encoded in these vertex-wise maps might also provide a way to investigate atypical topography in less explored cortical areas.

Acknowledgements. Data were provided by the Human Connectome Project, WU-Minn Consortium (Principal Investigators: David Van Essen and Kamil Ugurbil; 1U54MH091657) funded by the 16 NIH Institutes and Centers that support the NIH Blueprint for Neuroscience Research; and by the McDonnell Center for Systems Neuroscience at Washington University [31].

References

1. Abdollahi, R.O., et al.: Correspondences between retinotopic areas and myelin maps in human visual cortex. Neuroimage **99**, 509–524 (2014)
2. Bronstein, M.M., Bruna, J., LeCun, Y., Szlam, A., Vandergheynst, P.: Geometric deep learning: going beyond Euclidean data. IEEE Signal Process. Mag. **34**(4), 18–42 (2017)
3. Chen, H., Dou, Q., Yu, L., Qin, J., Heng, P.A.: Voxresnet: deep voxelwise residual networks for brain segmentation from 3D MR images. NeuroImage **170**, 446–455 (2018)
4. Defferrard, M., Bresson, X., Vandergheynst, P.: Convolutional neural networks on graphs with fast localized spectral filtering. arXiv preprint arXiv:1606.09375 (2016)
5. Desikan, R.S., et al.: An automated labeling system for subdividing the human cerebral cortex on MRI scans into gyral based regions of interest. Neuroimage **31**(3), 968–980 (2006)
6. Fey, M., Lenssen, J.E.: Fast graph representation learning with PyTorch Geometric. In: ICLR Workshop on Representation Learning on Graphs and Manifolds (2019)
7. Frost, M.A., Goebel, R.: Measuring structural-functional correspondence: spatial variability of specialised brain regions after macro-anatomical alignment. Neuroimage **59**(2), 1369–1381 (2012)
8. Gal, Y., Ghahramani, Z.: Dropout as a Bayesian approximation: representing model uncertainty in deep learning. In: International Conference on Machine Learning, pp. 1050–1059. PMLR (2016)
9. Given, N.A.: Benchmarking geometric deep learning for cortical segmentation and neurodevelopmental phenotype prediction (2021, in preparation)
10. Glasser, M.F., et al.: A multi-modal parcellation of human cerebral cortex. Nature **536**(7615), 171–178 (2016)
11. Glasser, M.F., et al.: The human connectome project's neuroimaging approach. Nat. Neurosci. **19**(9), 1175–1187 (2016)
12. Glasser, M.F., et al.: The minimal preprocessing pipelines for the human connectome project. Neuroimage **80**, 105–124 (2013)
13. Glasser, M.F., Van Essen, D.C.: Mapping human cortical areas in vivo based on myelin content as revealed by T1-and T2-weighted MRI. J. Neurosci. **31**(32), 11597–11616 (2011)
14. Gordon, E.M., et al.: Individual-specific features of brain systems identified with resting state functional correlations. Neuroimage **146**, 918–939 (2017)
15. Gratton, C., et al.: Defining individual-specific functional neuroanatomy for precision psychiatry. Biol. Psychiatr. **88**, 28-39 (2019)
16. Havaei, M., et al.: Brain tumor segmentation with deep neural networks. Med. Image Anal. **35**, 18–31 (2017)
17. Heckemann, R.A., Hajnal, J.V., Aljabar, P., Rueckert, D., Hammers, A.: Automatic anatomical brain MRI segmentation combining label propagation and decision fusion. NeuroImage **33**(1), 115–126 (2006)
18. Kipf, T.N., Welling, M.: Semi-supervised classification with graph convolutional networks. arXiv preprint arXiv:1600.02907 (2016)
19. Kong, R., et al.: Spatial topography of individual-specific cortical networks predicts human cognition, personality, and emotion. Cereb. Cortex **29**(6), 2533–2551 (2019)
20. Makropoulos, A., et al.: The developing human connectome project: a minimal processing pipeline for neonatal cortical surface reconstruction. Neuroimage **173**, 88–112 (2018)

21. Miller, K.L., et al.: Multimodal population brain imaging in the UK Biobank prospective epidemiological study. Nat. Neurosci. **19**(11), 1523–1536 (2016)
22. Monti, F., Boscaini, D., Masci, J., Rodola, E., Svoboda, J., Bronstein, M.M.: Geometric deep learning on graphs and manifolds using mixture model CNNs. In: Proceedings of the IEEE Conference on Computer Vision and Pattern Recognition (CVPR) (2017)
23. Öngür, D., Ferry, A.T., Price, J.L.: Architectonic subdivision of the human orbital and medial prefrontal cortex. J. Comp. Neurol. **460**(3), 425–449 (2003)
24. Reinke, A., et al.: Common limitations of image processing metrics: a picture story. arXiv preprint arXiv:2104.05642 (2021)
25. Robinson, E.C., et al.: Multimodal surface matching with higher-order smoothness constraints. Neuroimage **167**, 453–465 (2018)
26. Robinson, E.C., et al.: MSM: a new flexible framework for multimodal surface matching. Neuroimage **100**, 414–426 (2014)
27. Ronneberger, O., Fischer, P., Brox, T.: U-net: convolutional networks for biomedical image segmentation. In: Navab, N., Hornegger, J., Wells, W., Frangi, A. (eds.) MICCAI 2015. LNCS, vol. 9351, pp. 234–241. Springer, Cham (2015). https://doi.org/10.1007/978-3-319-24574-4_28
28. Sabuncu, M.R., Yeo, B.T., Van Leemput, K., Fischl, B., Golland, P.: A generative model for image segmentation based on label fusion. IEEE Trans. Med. Imaging **29**(10), 1714–1729 (2010)
29. Sudre, C.H., Li, W., Vercauteren, T., Ourselin, S., Cardoso, M.J.: Generalised dice overlap as a deep learning loss function for highly unbalanced segmentations. In: Cardoso, M., et al. (eds.) DLMIA 2017, ML-CDS 2017. LNCS, vol. 10553, pp. 240–248. Springer (2017). https://doi.org/10.1007/978-3-319-67558-9_28
30. Van Essen, D.C., Glasser, M.F., Dierker, D.L., Harwell, J., Coalson, T.: Parcellations and hemispheric asymmetries of human cerebral cortex analyzed on surface-based atlases. Cereb. Cortex **22**(10), 2241–2262 (2012)
31. Van Essen, D.C., et al.: The WU-Minn human connectome project: an overview. Neuroimage **80**, 62–79 (2013)
32. Zhao, F., et al.: Spherical u-net on cortical surfaces: methods and applications. In: Chung, A., Gee, J., Yushkevich, P., Bao, S. (eds.) IPMI 2019, LNCS, vol. 11492, pp. 855–866. Springer, Cham (2019). https://doi.org/10.1007/978-3-030-20351-1_67

Deep Stacking Networks for Conditional Nonlinear Granger Causal Modeling of fMRI Data

Kai-Cheng Chuang[1,2] (iD), Sreekrishna Ramakrishnapillai[2], Lydia Bazzano[3], and Owen T. Carmichael[2(✉)] (iD)

[1] Medical Physics Graduate Program, Louisiana State University, Baton Rouge, LA, USA
[2] Biomedical Imaging Center, Pennington Biomedical Research Center, Baton Rouge, LA, USA
{Kai.Chuang,Owen.Carmichael}@pbrc.edu
[3] Department of Epidemiology, Tulane School of Public Health and Tropical Medicine, New Orleans, LA, USA

Abstract. Conditional Granger causality, based on functional magnetic resonance imaging (fMRI) time series signals, is the quantification of how strongly brain activity in a certain source brain region contributes to brain activity in a target brain region, independent of the contributions of other source regions. Current methods to solve this problem are either unable to model nonlinear relationships between source and target signals, unable to efficiently quantify time lags in source-target relationships, or require ad hoc parameter settings and post hoc calculations to assess conditional Granger causality. This paper proposes the use of deep stacking networks, with dilated convolutional neural networks (CNNs) as component parts, to address these challenges. The dilated CNNs nonlinearly model the target signal as a function of source signals. Conditional Granger causality is assessed in terms of how much modeling fidelity increases when additional dilated CNNs are added to the model. Time lags between source and target signals are estimated by analyzing estimated dilated CNN parameters. Our technique successfully estimated conditional Granger causality, did not spuriously identify false causal relationships, and correctly estimated time lags when applied to synthetic datasets and data generated by the STANCE fMRI simulator. When applied to real-world task fMRI data from an epidemiological cohort, the method identified biologically plausible causal relationships among regions known to be task-engaged and provided new information about causal structure among sources and targets that traditional single-source causal modeling could not provide. The proposed method is promising for modeling complex Granger causal relationships within brain networks.

Keywords: Functional causal modeling (FCM) · Deep stacking networks (DSNs) · Functional magnetic resonance imaging (fMRI)

1 Introduction

Conditional Granger causal modeling has become an important concept in neuroscience as functional neuroimaging methods, including functional magnetic resonance imaging

© Springer Nature Switzerland AG 2021
A. Abdulkadir et al. (Eds.): MLCN 2021, LNCS 13001, pp. 113–124, 2021.
https://doi.org/10.1007/978-3-030-87586-2_12

(fMRI), enable the precise mapping of brain activity within neural circuitry underlying perception, cognition, and behavior [1, 2]. Conditional Granger causality (CGC) is defined as the unique influence that activity in one brain region (the "source") exerts over activity in another region (the "target") after accounting for the influence of activity in other source regions [3–5] (Fig. 1). CGC provides deeper information about brain circuit functioning than correlational concepts of functional connectivity by capturing directional relationships that underlie information flow during rest or task execution. Charting such directional relationships is important to the study of brain development, dysfunction during diseases, and degeneration during aging, as well as normal functioning during perception, cognition, behavior, and consciousness [2, 6–9].

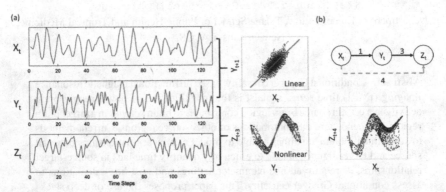

Fig. 1. (**a**) A hypothetical three time series dataset, where X_t has a strong linear relationship with Y_{t+1}, suggesting a causal relationship between X and Y with a time lag of 1 time step. Y_t has a nonlinear causal relationship with Z_t with a time lag of 3 time steps. (**b**) Conditional Granger causality analysis allows us to determine that while X and Z appear to have a nonlinear causal relationship with a time lag of 4 time steps, this relationship is really just a reflection of causal relationships with the intermediary source Y.

The multiple vector auto-regression (MVAR) model has been widely used to assess conditional Granger causality. This paper addresses two key challenges limiting the utility of the MVAR model. First, **nonlinear relationships** between source and target signals (caused by complex neural dynamics giving rise to those signals) are commonly studied in neuroscience but are not well modeled by the MVAR model which is inherently linear [10–16]. In addition, identifying **time lags** in MVAR is computationally complex, typically requiring estimation of all models representing all possible time lags followed by model scoring using the Akaike information criterion (AIC) or similar criteria [16–18]. Kernel Granger causality (KGC), extended Granger causality, recurrent neural network methods [19–21], and multilayer perceptron methods [22] solve the nonlinearity problem for traditional single-source causality but do not address conditional Granger causality.

Nauta et al. proposed the Temporal Causal Discovery Framework (TCDF), an attention-based dilated convolutional neural network (CNN) that solves the Granger causality problem efficiently while allowing fast estimation of time lags via interpretation of dilated CNN internal parameters [23]. However, the resulting causalities depend on operating parameters—thresholds on attention scores and a loss function—that may be difficult to set.

We propose a machine learning algorithm called deep stacking networks (DSNs), which improves upon CGC (Fig. 2(a)) by using dilated CNNs to estimate nonlinear relationships with efficient time lag estimation (Fig. 2(b)). Computational efficiency of training is an additional benefit of DSNs, as each dilated CNN can be trained separately from the others and some can be trained in parallel.

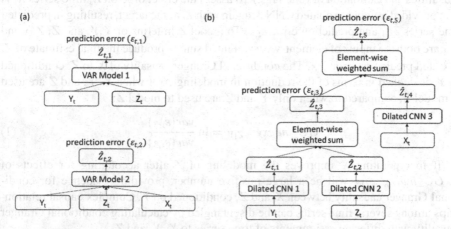

Fig. 2. (a) MVAR assesses X as a source for the target Z, conditioned on Y $(X \rightarrow Z|Y)$ by comparing the prediction errors of VAR models that use only Y and Z, vs. X, Y, and Z simultaneously. (b) A deep stacking network solves this same problem using a series of CNNs, providing modeling of nonlinear causal relationships and efficient computation of time lags.

2 Materials and Methods

2.1 Deep Stacking Network

The philosophy of DSN design is based on the concept of stacking proposed by Wolpert where simple modules of functions or classifiers are trained first, and then they are "stacked" on top of each other to compose complex functions or classifiers [24]. Deng et al. presented the basic form of the DSN architecture that consists of many stacking modules, each of which takes a simplified form of shallow multilayer perceptron using convex optimization for learning perceptron weights [25, 26]. The 'stacking' architecture is built by concatenating the output from previous modules with the original input to form the new "input" for the next module. Each layer of hidden units learns to represent features that capture higher order correlations within the input data. One benefit of DSNs is scalability, owing to their ability to train modules in parallel [25, 26].

2.2 Conditional Nonlinear Granger Causal Modeling with DSN

Following the above example of sources X and Y, and target Z, we train a DSN with individual dilated CNN to transform X_t, Y_t, and Z_t into Z_t, resulting in Z_t estimates $\hat{Z}_{t,1}$,

$\hat{Z}_{t,2}, \ldots, \hat{Z}_{t,5}$ (See Algorithm 1 and Fig. 2(b)). First, to evaluate the effect of Y and Z on Z, dilated CNNs 1 and 2 are trained respectively to transform Y_t and Z_t into Z_t, resulting in estimates $\hat{Z}_{t,1}$ and $\hat{Z}_{t,2}$ and prediction errors $\varepsilon_{t,1}$ and $\varepsilon_{t,2}$. Then, to represent the best estimate of Z based on both of Y and Z, $\hat{Z}_{t,1}$ and $\hat{Z}_{t,2}$ both provide as inputs to the third module, which estimates an element-wise weighted sum of the inputs to predict Z_t, resulting in estimate $\hat{Z}_{t,3}$ and prediction error of $\varepsilon_{t,3}$. (Element-wise weighted sum is used instead of another dilated CNN as dilated CNN complicates the assessment of causalities and estimation of time lags.) To assess the effect of X on Z, time series X_t is then provided as input to dilated CNN 3, again with Z_t as the target, resulting in predicted time series $\hat{Z}_{t,4}$ and prediction error $\varepsilon_{t,4}$. To model Z in terms of X, Y, and Z, $\hat{Z}_{t,3}$ and $\hat{Z}_{t,4}$ are both the inputs of element-wise weighted sum to produce the final estimate of Z, $\hat{Z}_{t,5}$, and prediction error $\varepsilon_{t,5}$. The conditional Granger causality of X to Z, conditioned on Y, is defined in terms of the reduction in modeling error when X, Y, and Z are used to model Z, compared to when only Y and Z are used to model Z:

$$GC\_index_{X \to Z|Y} = \ln \frac{var(\varepsilon_{t,3})}{var(\varepsilon_{t,5})} \tag{1}$$

If incorporating X improves the modeling of Z after accounting for effects of Y, $GC\_index_{X \to Z|Y}$ will be a large positive number, providing evidence for conditional Granger causality between X and Z, conditioned on Y. Complex causal relationships among several time series can be disentangled by calculating conditional Granger causality with differing assignments of time series to X, Y, and Z.

Algorithm 1: Conditional nonlinear Granger causal modeling with DSN

DSN module (input(s), target, estimate, prediction error)

Step 1. Train Dilated CNN 1 (Y_t, Z_t, $\hat{Z}_{t,1}$, $\varepsilon_{t,1}$) and Dilated CNN 2 (Z_t, Z_t, $\hat{Z}_{t,2}$, $\varepsilon_{t,2}$)

Step 2. Estimate Element-wise weighted sum ([$\hat{Z}_{t,1}$, $\hat{Z}_{t,2}$], Z_t, $\hat{Z}_{t,3}$, $\varepsilon_{t,3}$)

Step 3. Train Dilated CNN 3 (X_t, Z_t, $\hat{Z}_{t,4}$, $\varepsilon_{t,4}$)

Step 4. Estimate Element-wise weighted sum ([$\hat{Z}_{t,3}$, $\hat{Z}_{t,4}$], Z_t, $\hat{Z}_{t,5}$, $\varepsilon_{t,5}$)

Step 5. Calculate $GC\_index_{X \to Z|Y} = \ln \frac{var(\varepsilon_{t,3})}{var(\varepsilon_{t,5})}$

Nonlinear parametric ReLU (PReLU) activation functions were used in each hidden layer to reveal linear and nonlinear relationships between source and target regions. Each dilated CNN had 3 hidden layers, a 1D kernel, a kernel size of 2, and zero causal padding, inspired by the well-known WaveNet architecture [23, 27]. A dilated CNN is a convolution where the kernel is applied over a length larger than its length by skipping input values with a certain step. Zero causal padding was used to prevent future time point values from influencing the prediction of current time points. To efficiently estimate time lags in source-target relationships, we interpreted the kernel weights W_i in each dilated CNN.

2.3 Model Validation and Application

We applied the proposed method to synthetic time series data, simulated fMRI data from a public-domain simulator, and a real-world fMRI dataset. We assessed how well true

Granger causal relationships and time lags programmed into the synthetic and simulated data were identified by our method, as well as whether spurious causal relationships were identified. Also, the identified Granger causalities from the proposed method were compared with those from previous methods, MVAR and TCDF. For each synthetic and simulated dataset, $100\,X, Y, Z$ time series triples were generated. Seventy-two partici-pants were included in the real-world fMRI dataset. Ten-fold cross validation was used to repeatedly train the DSNs and quantify causal relationships within the testing data. The mean and standard deviation of the GC index across cross validation folds is pre-sented. We consider the evidence for a particular conditional Granger causal relationship strong when the corresponding GC index is statistically significantly greater than 0 in one-tailed student's t-test (p-value < 0.05). Also, the identified Granger causalities from the proposed method were compared with those from previous methods, MVAR [16] and TCDF [23]. The same statistical significance criterion (p-value < 0.05) has been applied for MVAR. TCDF analysis followed the same dilated CNNs architecture and statistical criterion as Nauta et al. [23].

3 Experiments and Results

3.1 Synthetic Dataset

Synthetic Dataset 1: Linear Causal Relationships. Each time series had 128 time points generated according to the following formulas [28], where $w_{t,1}, w_{t,2}, w_{t,3}$ were Gaussian white noise with standard deviation of 0.1. True causal relationships within this dataset are represented in Fig. 3(a).

$$X_t = 0.95 \times \sqrt{2} \times X_{t-1} - 0.9025 \times X_{t-2} + w_{t,1}$$

$$Y_t = -0.5 \times X_{t-1} + w_{t,2} \tag{2}$$

$$Z_t = -0.5 \times Y_{t-1} + 0.5 \times Z_{t-1} + w_{t,3}$$

Synthetic Dataset 2: Nonlinear Causal Relationships. Time series X_t and Y_t share the same common Gaussian white noise, $w_{t,1}$, with standard deviation of 0.1. Independent Gaussian white noise, $w_{t,2}$, with standard deviation of 0.1, was added to time series Z_t. The length of the time series was set to 128 time points. The causal relationships of the set are shown in Fig. 3(B).

$$X_t = 0.95 \times \sqrt{2} \times X_{t-1} - 0.9025 \times X_{t-2} - 0.9 \times Y_{t-1} + 0.5 \times w_{t,1}$$

$$Y_t = -1.05 \times Y_{t-1} - 0.85 \times Y_{t-2} - 0.8 \times X_{t-1} + 0.5 \times w_{t,1} \tag{3}$$

$$Z_t = 0.1 + 0.4 \times Z_{t-2} + \frac{2.4 - 0.9 \times X_{t-3}}{1 + e^{-4 \times X_{t-3}}} + w_{t,2}$$

Results. The proposed method correctly identified true conditional Granger causalities $X \rightarrow Y|Z$ and $Y \rightarrow Z|X$ with statistically significant GC indices in synthetic dataset 1 (Eq. (2)). In addition, a causal relationship such as $X \rightarrow Z$ that would be viewed as significant when X and Z are viewed in isolation was non-significant when the influence of the intermediary source (Y) was considered ($X \rightarrow Z|Y$). Similarly, true conditional Granger causalities in synthetic dataset 2 were identified. The method also accurately identified the true time lags between all sources and targets ($Y \rightarrow X|Z$). Moreover, the magnitude of the GC index also reflected the strength of the causal relationship in synthetic time series 1. Specifically, there was strong evidence for $X \rightarrow Y|Z$ ($GC_{index\,X\rightarrow Y|Z} = 0.3151$), reflecting that Y depended solely on the previous time steps of X and not other time series (Eq. (3)). In contrast, Z depended on previous time steps of both Y and Z, suggesting less of a unique contribution of one or the other; this was reflected in a lower GC index (0.0494). The strength of causalities in synthetic time series 2 are difficult to interpret using the equations themselves with mutual and nonlinear causal relationships. However, the GC indices indicated a strong causality of $X \rightarrow Y|Z$, a moderate causality of $X \rightarrow Z|Y$, and a weak causality of $Y \rightarrow X|Z$. Figure 4 shows the comparison result of synthetic datasets 1 and 2 for three conditional Granger causality methods. Our proposed method, DSN-CNNs, fully detected the causal relationships and time lags which were already given in the datasets. However, the models of MVAR and TCDF did not perform well (Fig. 4).

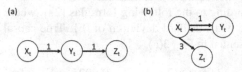

Fig. 3. (a) Synthetic dataset 1: linear case with causalities of $X \rightarrow Y$ and $Y \rightarrow Z$, and self-causalities of X and Z. (b) Synthetic dataset 2: nonlinear case with mutual causalities between X and Y, and a flow of $X \rightarrow Z$, and self-causalities of $X, Y,$ and Z.

3.2 Simulated fMRI Dataset

The STANCE simulator, developed by Hill et al., was used to simulate task-based fMRI data [29]. For simulated fMRI dataset, the event designs were produced with causal relationships among time series then convolved with a canonical hemodynamic response function (HRF) followed by the addition of simulated system and physiological noise, of which the magnitude was about 10% of the simulated fMRI signal. Each time series had 128 time points.

Simulated Dataset 1. Source time series X_t and Y_t were generated with 26 randomly spaced events, each with a duration of 1 time step. The target time series, Z_t, would have an event with 70% probability if there was an event at X_{t-2} and would have an event with 30% probability if there was an event at Y_{t-5}. The causal relationships of this dataset are represented in Fig. 5(a).

Table 1. The identified causal relationships and time lags between time series, X, Y, Z, in synthetic dataset 1 & 2. The causalities programmed into the datasets a priori are marked in bold.

Causality	Synthetic dataset 1			Causality	Synthetic dataset 2		
	GC index	p-value	Time lags		GC index	p-value	Time lags
$X \to Y\mid Z$	0.3151 ± 0.0778	<0.0001	1	$X \to Y\mid Z$	0.2666 ± 0.1701	0.0006	1
$Z \to Y\mid X$	-0.0016 ± 0.0019	0.9836	-	$Z \to Y\mid X$	-0.0005 ± 0.0012	0.9025	-
$X \to Z\mid Y$	0.0164 ± 0.0309	0.0730	-	$X \to Z\mid Y$	0.0707 ± 0.0115	0.0292	3
$Y \to Z\mid X$	0.0494 ± 0.0220	<0.0001	1	$Y \to Z\mid X$	-0.0001 ± 0.0035	0.5189	-
$Y \to X\mid Z$	0.0078 ± 0.0283	0.2159	-	$Y \to X\mid Z$	0.0056 ± 0.0087	0.0299	1
$Z \to X\mid Y$	-0.0016 ± 0.0015	0.9906	-	$Z \to X\mid Y$	-0.0004 ± 0.0017	0.7354	-

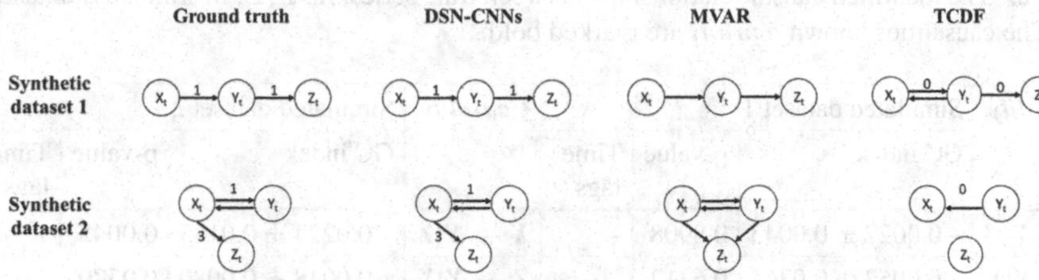

Fig. 4. Comparison of the three CGC methods with synthetic datasets.

Simulated Dataset 2. Source time series X_t was generated with 26 randomly spaced events, each with a duration of 1 time step. Source time series Y_t had an event with 90% probability if there was an event at X_{t-1}. The target time series Z_t had an event with 50% probability if there was an event at X_{t-1} and had an event with 50% probability if there was an event at Y_{t-1}. The causal relationships of this dataset are shown in Fig. 5(b).

Result. The true conditional Granger causalities of $X \to Z\mid Y$ and $Y \to Z\mid X$ in simulated dataset 1 have been identified by the proposed method with statistically significant GC indices, and the correct time lags were identified (Table 2). The proposed method also revealed the true conditional Granger causalities in simulated dataset 2 and accurately estimated time lags. None of the spurious conditional Granger causalities had GC index values that were significantly larger than 0. Again, our proposed method, DSN-CNNs, outperformed the other two methods (Fig. 6).

3.3 Real-World fMRI Dataset

We applied DSN-CNNs to task fMRI data collected from the Bogalusa Heart Study, a community-based cohort study of risk factors for cardiovascular disease in a biracial population of children in a semi-rural town in southeastern Louisiana. The subjects performed the Stroop task during fMRI scan [30]. In this paper, 82 participants were

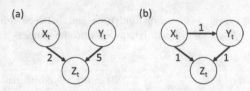

Fig. 5. (a) Simulated dataset 1: simulated fMRI dataset with causalities of $X \rightarrow Z$ and $Y \rightarrow Z$. (b) Simulated dataset 2: simulated fMRI dataset with causalities $X \rightarrow Y$, $X \rightarrow Z$, and $Y \rightarrow Z$.

Table 2. The identified causal relationships between time series, X, Y, Z, in simulated dataset 1 & 2. The causalities known *a priori* are marked bold.

Causality	Simulated dataset 1			*Causality*	Simulated dataset 2		
	GC index	p-value	Time lags		GC index	p-value	Time lags
$X \rightarrow Y|Z$	-0.0027 ± 0.0043	0.8908	-	$X \rightarrow Y|Z$	0.0224 ± 0.0145	0.0045	1
$Z \rightarrow Y|X$	-0.0053 ± 0.0367	0.6312	-	$Z \rightarrow Y|X$	-0.0048 ± 0.0080	0.9379	-
$X \rightarrow Z|Y$	0.0238 ± 0.0115	0.0119	2	$X \rightarrow Z|Y$	0.0394 ± 0.0115	0.0012	1
$Y \rightarrow Z|X$	0.0194 ± 0.0175	0.0070	5	$Y \rightarrow Z|X$	0.0041 ± 0.0018	0.0018	1
$Y \rightarrow X|Z$	-0.0015 ± 0.0012	0.9958	-	$Y \rightarrow X|Z$	0.0001 ± 0.0071	0.4802	-
$Z \rightarrow X|Y$	-0.0018 ± 0.0021	0.9974	-	$Z \rightarrow X|Y$	-0.0004 ± 0.0052	0.5975	-

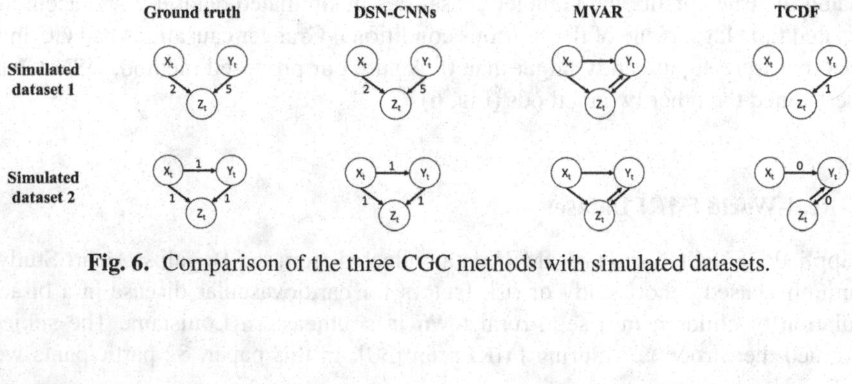

Fig. 6. Comparison of the three CGC methods with simulated datasets.

included; however, 10 subjects with significant head motion during the fMRI scan were excluded from the analysis. Brain MRI scans were performed on a GE Discovery 3T scanner at Pennington Biomedical Research Center. T1-weighted structural MPRAGE (voxel size, $1 \times 1 \times 1$ mm^3; voxel array, $256 \times 256 \times 176$; flip angle, 8°; NEX, 1) and axial 2D gradient echo EPI BOLD fMRI (voxel size, $3.5 \times 3.5 \times 3.5$ mm^3; voxel array, $64 \times 64 \times 44$; flip angle, 90°; TE, 30 ms; TR, 3000 ms; NEX, 1) were acquired. Two hundred volumes were acquired over the course of task execution. Preprocessing of fMRI included slice timing correction, head motion correction, smoothing, co-registration to the T1-weighted image, and warping of T1-weighted data to a standard coordinate frame (using Statistical Parametric Mapping 12). Cardiac and respiratory time series were regressed out of the data using RETROICOR [31]. The 3D locations of regions of interest (ROIs) previously identified as activated by the Stroop task (fusiform gyrus, occipital gyrus, precuneus, and thalamus) were extracted from Sheu et al. [32], and a single summary fMRI time series was extracted from each ROI in each scan by in-house MATLAB code. DSN-CNNs was applied to all possible assignments of ROIs to the roles of X_t, Y_t, and Z_t to explore conditional Granger causal relationships in the real dataset.

Result. Table 3 shows the conditional Granger causalities with GC indices that were statistically significantly larger than 0. Occipital gyrus was seen as a significant source for the target thalamus region, independent of influences from the fusiform gyrus and precuneus. Conversely, the thalamus was considered a significant source for the occipital gyrus, independent of influences from the fusiform gyrus and precuneus. The strong directional relationships between the occipital gyrus and thalamus have biological plausibility. In the visual system, the lateral geniculate nucleus (LGN) of the thalamus is the gateway through which visual information reaches the visual cortex located in the occipital gyrus; neither the precuneus nor the fusiform gyrus are intermediary regions along this pathway [33, 34]. In fact, our method correctly determined that causal relationships between a fusiform source region and thalamus and occipital target regions was actually well accounted for by the effects of intermediary source regions on the target (Table 4). This example illustrates the power of conditional Granger causal modeling: it can prevent identification of spurious causal relationships by accounting for intermediary sources [35–38].

Table 3. The identified causal relationships between ROIs in real-world fMRI dataset.

Causality	GC index	p-value	Time lags
Occipital → Thalamus|Fusiform	0.0173 ± 0.0232	0.0262	1
Occipital → Thalamus|Precuneus	0.0323 ± 0.0254	0.0035	1
Thalamus → Occipital|Fusiform	0.0213 ± 0.0181	0.0032	0
Thalamus → Occipital|Precuneus	0.0180 ± 0.0225	0.0466	0

Table 4. The comparison of causality between pairwise analysis and proposed method.

Causality	GC index	p-value
Fusiform → Thalamus	0.0280 ± 0.0234	0.0029
Fusiform → Occipital	0.0110 ± 0.0154	0.0394
Fusiform → Thalamus|Occipital	-0.0129 ± 0.0296	0.2745
Fusiform → Occipital|Thalamus	0.0006 ± 0.0296	0.8043

4 Discussion

The proposed method showed promising results when applied to a synthetic time series and simulated fMRI dataset compared to previous methods. Figure 4 and Fig. 6 showed that the proposed method, DSN-CNNs, was accurate for both linear and nonlinear relationship detection. MVAR successfully estimated linear causality in synthetic dataset 1, but it identified spurious causalities in the nonlinear case (synthetic dataset 2 and simulated datasets 1 and 2). Direct time lag estimation is not possible when using MVAR, as time lag estimation requires comparing all models representing all possible time lags. Identifying conditional Granger causality by TCDF for both linear and nonlinear datasets was difficult, as the resulting causalities depend on an operating parameter—a threshold on attention scores—that may lead to the spurious causalities.

5 Conclusion

We proposed a DSN architecture that characterizes conditional nonlinear Granger causal relationships among multiple time series while efficiently estimating time lags in those relationships. When applied to synthetic time series and simulated fMRI data, the method correctly identified causal relationships and time lags that were programmed into the data *a priori*, with no identification of spurious causal relationships. When applied to real-world data, biologically plausible conditional Granger causal relationships were identified. Future work should extend this approach to causal relationship discovery within large-scale brain networks [23, 36, 39].

Acknowledgments. Funding for this work was provided by NIH grants R01AG041200 and R01AG062309 as well as the Pennington Biomedical Research Foundation.

References

1. Friston, K., Frith, C., Frackowiak, R.: Time-dependent changes in effective connectivity measured with PET. Hum. Brain Mapp. **1**(1), 69–79 (1993)
2. Seth, A.K., Barrett, A.B., Barnett, L.: Granger causality analysis in neuroscience and neuroimaging. J. Neurosci. **35**(8), 3293–3297 (2015)
3. Chen, Y., et al.: Analyzing multiple nonlinear time series with extended Granger causality. Phys. Lett. A **324**(1), 26–35 (2004)

4. Zhou, Z., et al.: Analyzing brain networks with PCA and conditional Granger causality. Hum. Brain Mapp. **30**(7), 2197–2206 (2009)
5. Zhou, Z., et al.: A conditional Granger causality model approach for group analysis in functional magnetic resonance imaging. Magn. Reson. Imaging **29**(3), 418–433 (2011)
6. Dai, W., et al.: Mild cognitive impairment and Alzheimer disease: patterns of altered cerebral blood flow at MR imaging. Radiology **250**(3), 856–866 (2009)
7. Goebel, R., et al.: Investigating directed cortical interactions in time-resolved fMRI data using vector autoregressive modeling and Granger causality mapping. Magn. Reson. Imaging **21**(10), 1251–1261 (2003)
8. He, J., et al.: Influence of functional connectivity and structural MRI measures on episodic memory. Neurobiol. Aging **33**(11), 2612–2620 (2012)
9. Logothetis, N.K., et al.: Neurophysiological investigation of the basis of the fMRI signal. Nature **412**(6843), 150–157 (2001)
10. Aertsen, A., et al.: Dynamics of neuronal firing correlation: modulation of "effective connectivity." J. Neurophysiol. **61**(5), 900–917 (1989)
11. Buxton, R.B., et al.: Modeling the hemodynamic response to brain activation. Neuroimage **23**, S220–S233 (2004)
12. Grosmark, A.D., Buzsáki, G.: Diversity in neural firing dynamics supports both rigid and learned hippocampal sequences. Science **351**(6280), 1440–1443 (2016)
13. Johnston, L.A., et al.: Nonlinear estimation of the BOLD signal. Neuroimage **40**(2), 504–514 (2008)
14. Liao, W., et al.: Kernel Granger causality mapping effective connectivity on fMRI data. IEEE Trans. Med. Imaging **28**(11), 1825–1835 (2009)
15. Marinazzo, D., et al.: Nonlinear connectivity by Granger causality. Neuroimage **58**(2), 330–338 (2011)
16. Barnett, L., Seth, A.K.: The MVGC multivariate Granger causality toolbox: a new approach to Granger-causal inference. J. Neurosci. Methods **223**, 50–68 (2014)
17. Guo, H., et al.: Kernel Granger causality based on back propagation neural network fuzzy inference system on fMRI data. IEEE Trans. Neural Syst. Rehabil. Eng. **28**(5), 1049–1058 (2020)
18. Li, F., et al.: Unified model selection approach based on minimum description length principle in Granger causality analysis. IEEE Access **8**, 68400–68416 (2020)
19. Chivukula, A.S., Li, J., Liu, W.: Discovering Granger-causal features from deep learning networks. In: Mitrovic, T., Xue, B., Li, X. (eds.) AI 2018. LNCS, vol. 11320, pp. 692–705. Springer, Cham (2018). https://doi.org/10.1007/978-3-030-03991-2_62
20. Duggento, A., Guerrisi, M., Toschi, N.: Echo State Network models for nonlinear Granger causality. bioRxiv, p. 651679 (2019)
21. Guo, T., Lin, T., Lu, Y.: An interpretable LSTM neural network for autoregressive exogenous model. arXiv preprint arXiv:1804.05251 (2018)
22. Tank, A., et al.: Neural granger causality for nonlinear time series. arXiv preprint arXiv:1802.05842 (2018)
23. Nauta, M., Bucur, D., Seifert, C.: Causal discovery with attention-based convolutional neural networks. Mach. Learn. Knowl. Extr. **1**(1), 312–340 (2019)
24. Wolpert, D.H.: Stacked generalization. Neural Netw. **5**(2), 241–259 (1992)
25. Deng, L., Hutchinson, B., Yu, D.: Parallel training for deep stacking networks. In: Thirteenth Annual Conference of the International Speech Communication Association (2012)
26. Deng, L. Yu, D.: Deep convex net: a scalable architecture for speech pattern classification. In: Twelfth Annual Conference of the International Speech Communication Association (2011)
27. Van den Oord, A., et al.: Wavenet: A generative model for raw audio. arXiv preprint arXiv:1609.03499 (2016)

28. Gourévitch, B., Le Bouquin-Jeannès, R., Faucon, G.: Linear and nonlinear causality between signals: methods, examples and neurophysiological applications. Biol. Cybern. **95**(4), 349–369 (2006)
29. Hill, J.E., et al. A task-related and resting state realistic fMRI simulator for fMRI data validation. In: Medical Imaging 2017: Image Processing. 2017. International Society for Optics and Photonics (2017)
30. Carmichael, O., et al.: High-normal adolescent fasting plasma glucose is associated with poorer midlife brain health: Bogalusa Heart Study. J. Clin. Endocrinol. Metab. **104**(10), 4492–4500 (2019)
31. Glover, G.H., Li, T.Q., Ress, D.: Image-based method for retrospective correction of physiological motion effects in fMRI: RETROICOR. Magn. Reson. Med.: Off. J. Int. Soc. Magn. Reson. Med. **44**(1), 162–167 (2000)
32. Sheu, L.K., Jennings, J.R., Gianaros, P.J.: Test–retest reliability of an fMRI paradigm for studies of cardiovascular reactivity. Psychophysiology **49**(7), 873–884 (2012)
33. Guido, W.: Development, form, and function of the mouse visual thalamus. J. Neurophysiol. **120**(1), 211–225 (2018)
34. Usrey, W.M., Alitto, H.J.: Visual functions of the thalamus. Ann. Rev. Vis. Sci. **1**, 351–371 (2015)
35. Roebroeck, A., Formisano, E., Goebel, R.: Mapping directed influence over the brain using Granger causality and fMRI. Neuroimage **25**(1), 230–242 (2005)
36. Wang, X., et al.: Large-scale Granger causal brain network based on resting-state fMRI data. Neuroscience **425**, 169–180 (2020)
37. Blinowska, K.J., Kuś, R., Kamiński, M.: Granger causality and information flow in multivariate processes. Phys. Rev. E **70**(5), 050902 (2004)
38. Roebroeck, A., Formisano, E., Goebel, R.: The identification of interacting networks in the brain using fMRI: model selection, causality and deconvolution. Neuroimage **58**(2), 296–302 (2011)
39. Deshpande, G., et al.: Multivariate Granger causality analysis of fMRI data. Hum. Brain Mapp. **30**(4), 1361–1373 (2009)

Dynamic Adaptive Spatio-Temporal Graph Convolution for fMRI Modelling

Ahmed El-Gazzar[✉], Rajat Mani Thomas, and Guido van Wingen

Department of Psychiatry, Amsterdam UMC, University of Amsterdam,
Amsterdam, The Netherlands

Abstract. The characterisation of the brain as a functional network in
which the connections between brain regions are represented by correlation values across time series has been very popular in the last years.
Although this representation has advanced our understanding of brain
function, it represents a simplified model of brain connectivity that has a
complex dynamic spatio-temporal nature. Oversimplification of the data
may hinder the merits of applying advanced non-linear feature extraction
algorithms. To this end, we propose a dynamic adaptive spatio-temporal
graph convolution (DAST-GCN) model to overcome the shortcomings of
pre-defined static correlation-based graph structures. The proposed approach allows end-to-end inference of dynamic connections between brain
regions via layer-wise graph structure learning module while mapping
brain connectivity to a phenotype in a supervised learning framework.
This leverages the computational power of the model, data and targets
to represent brain connectivity, and could enable the identification of
potential biomarkers for the supervised target in question. We evaluate
our pipeline on the UKBiobank dataset for age and gender classification
tasks from resting-state functional scans and show that it outperforms
currently adapted linear and non-linear methods in neuroimaging. Further, we assess the generalizability of the inferred graph structure by
transferring the pre-trained graph to an independent dataset for the same
task. Our results demonstrate the task-robustness of the graph against
different scanning parameters and demographics.

Keywords: Functional connectivity · Spatio-temporal graph
convolution · Adaptive graph structure learning · UKBiobank

1 Introduction

A major goal of neuroimaging studies is to develop predictive models to analyze the relationship between whole brain functional connectivity patterns and
behavioural traits in order to better characterize brain function and dysfunction. Functional magnetic resonance imaging (fMRI) offers a promising window
to investigate the brain activity by measuring the blood oxygenation level which
offers a promising proxy to measure the neural activity of the brain. However,

© Springer Nature Switzerland AG 2021
A. Abdulkadir et al. (Eds.): MLCN 2021, LNCS 13001, pp. 125–134, 2021.
https://doi.org/10.1007/978-3-030-87586-2_13

extracting the spatiotemporal features from fMRI scans is challenging due to the high dimensionality of the data (∼1 M voxels), low signal-to-noise ratio and limited availability of labeled datasets. A standard approach to overcome these challenges is modelling functional connectivity as the pairwise correlations of the time-series of pre-defined brain regions and then feeding these to a classifier either as input features directly [19] or as the underlying structure of a graph [15]. The study of the brain as a graph offers a natural representation of the underlying mechanisms of brain activity [17]. Together with the application of graph theory approaches, the field of network neuroscience has significantly deepened our understanding of brain function and dysfunction. Recently, with the progress of geometric deep learning, graph convolution networks (GCNs) are being exploited in the analysis of fMRI scans [20,25]. A more befitting model for the dynamics of the brain are spatio-temporal GCNs (ST-GCNs) [24]. [2,7] recently evaluated the application of ST-GCNs for fMRI analysis for age and gender classification. While these adaptions show a promising direction for modelling the high dimensional signal of the fMRI scan using a spatio-temporal model, the result metrics do not display a significant improvement from shallow non-linear or linear models applied on the flattened correlation matrix. These findings align with the rise of multiple research papers that debate the merits of applying advance deep learning techniques in neuroimaging for phenotype prediction even with scaling of the number of training samples [9,16]. We hypothesize that these results are attributed to (1) the use of engineered features such as time-series correlations to represent the data of functional parcellations of brain regions, which limits the modelling capability of the deep learning techniques even if a suitable network architecture is implemented. (2) A static graph structure assumes static brain connectivity which contradicts the dynamic nature of brain connectivity [1] and its role in mapping to behavioural traits. (3) The adaption of exact neural network architectures that have shown previous success in other domains however are not well tailored to the nature of fMRI datasets such as the low temporal resolution and the scarcity of labelled samples. To address these issues, we propose a novel spatio-temporal graph convolution architecture with an adaptive graph structure learning. Our approach captures the spatio-temporal dependencies by combining graph convolutions and dilated 1D convolutions. Simultaneously, the graph structure learning module enables the inference of underlying dynamic graph structure. We evaluate our framework on the UKBiobank dataset for the tasks of sex and age classification form resting-state fMRI scans. We compare our results against different baselines as well as ablated versions of the model and we assess the robustness of the trained graph structure for the same task across an independent dataset. The remainder of this paper is ordered as follow: Sect. 2 describes the key components and the framework of the proposed architecture. Section 3 demonstrates the experiments conducted for the validation of the framework. Section 4 addresses generalizability of the inferred graph structures. Section 5 highlights the limitations of our work. Finally, Sect. 6 concludes this work and discusses the takeaways and future directions.

2   Methodology

2.1   Preliminaries

We model the brain as M directed dense graphs $\{G^k = (V, A^k)\}_{k=1}^{M}$, V are the nodes of the graph, where each node represent an anatomical region of the brain pre-defined using a template parcellation; $|V| = N$ is the number of regions in the template. $A^k \in R^{N \times N}$ represent a directed dense adjacency matrix initialized randomly. The signal on the graph is defined as $X \in R^{N \times T \times C}$ where T is the number of sampled data-points from the scan and C is the number of channels.

2.2   Temporal Lag Correction

The low temporal resolution and the existence of the temporal lag patterns in the fMRI signals was highlighted in previous studies [12]. This behaviour can limit the model ability to capture the underlying spatio-temporal signal. To correct for this latency, we compute the first and second derivative of the global time-course signal of each brain region, stack them as 3 channels and apply a 1×1 convolution with a linear activation function before passing the signal to the spatio-temporal blocks.

2.3   Temporal Feature Extraction

We adopt dilated 1D-convolution [13] as our temporal convolution layer (TCN) to capture a node's temporal trends. Unlike [13] however, we opt for non-causal convolutions, because (i) our objective is learning global temporal representations for a classification task, rather than a generative or horizon prediction task, and (ii) fMRI signals are known to exhibit a temporal lag structure so access to future timepoints could be essential to effectively model intrinsic brain networks. We use gated activation similar to [14]

$$z = \tanh(W_{k,f} * x) \odot \sigma(W_{k,g} * x) \tag{1}$$

where $*$ denotes a convolution operator, $\odot$ denotes an element-wise multiplication operator, σ is a sigmoid function, k is the layer index, f and g denote filter and gate, respectively, and W_f and W_g are the learnable filter and gate 1D convolutions. We empirically evaluate gated activation against rectified linear unites. We find that gated activation works better for modelling fMRI signal in terms of final metrics and loss convergence.

2.4   Spatial Feature Extraction

To extract the spatial correlations of data with unstructured topologies, extending neural networks to process graph-structured data has attracted widespread attention [21]. A series of studies has extended traditional convolution to model arbitrary graphs on spectral [4,5] or spatial [8,11] domains. Spectral-based methods use a graph spectral filter to smooth the input signals of nodes. Spatial-based

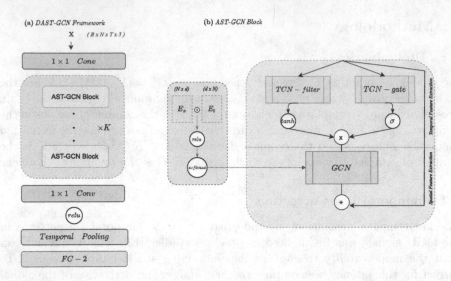

Fig. 1. Overview of the DAST-GCN architecture and its main components.

methods extract high-level representations of nodes by gathering feature information of neighbors. [11] shows that graph convolution operation can be well-approximated by 1-st order Chebyshev polynomial expansion and generalized to high-dimensional GCN as:

$$H = A'XW \tag{2}$$

where A' is the normalized adjacency matrix with self-loops. In neuroimaging, the most common approach to pre-define A is a correlation metric between the timecourses. In this work, we inspire from previous studies on adaptive adjacency matrix learning [3,22] and propose layer-wise graph structure learning module to capture the dynamic spatial dependency across different receptive fields. The normalized adjacency matrix can then be defined as:

$$A'_k = I_N + \text{Softmax}(\text{ReLU}(E_{s,k}E_{t,k})) \tag{3}$$

$E_s \in R^{N \times d}$ and $E_t \in R^{d \times N}$ are trainable source and target dictionaries, where d is a model hyper-parameter. By multiplying E_s and E_t, we derive the spatial dependency weights between the source nodes and the target nodes. We use the *ReLU* and *Softmax* activation functions to eliminate weak connections and normalize the adjacency matrix respectively. Finally, the identity matrix I_N is added for the self loops.

2.5   Framework of the Model

We present the framework of DAST-GCN in Fig. 1. First a 1×1 convolution layer with shared weights is applied on each node signal independently to scale up the number of features. Followed by a stack of spatio-temporal blocks. A spatial-temporal block is constructed by a gated temporal convolution layer (Gated

TCN) with shared weights across the nodes, an Adaptive graph convolution layer (GCN) and a residual connection. By stacking multiple spatial-temporal layers, DAST-GCN is able to handle spatial dependencies at different temporal levels. Subsequently, Another 1×1 convolution is applied to reduce the node feature dimensions to a single temporal vector, followed by a temporal pooling layer and a dropout layer for regularization. Finally a fully connected layer with softmax activation is applied for the binary classification task.

3 Experiments

3.1 Dataset

We evaluated our pipeline on rs-fMRI scans of 6709 (3255 males/3454 females, Age $= 67.3 \pm 7.2$) participants of UKBiobank dataset. First, the provided pre-processed scans were registered to an MNI template followed by applying the AAL atlas [18] to parcellate the brain into $N = 116$ anatomical regions of interest (ROIs). The time-courses (TR $= 0.7$ s, T $= 490$ timepoints) were averaged across the voxels of each ROI to obtain the signal for each node. We conducted binary sex classification and binary age classification for people above and under the age of 70 years to benchmark our framework.

3.2 Experimental Setup

We optimized the model for 200 epochs using Adam optimizer and binary cross-entropy loss with an initial learning rate of 0.001 and cosine scheduler with warm-up. A 5-fold cross-validation scheme was adapted for all the experiments. The models were trained on a Nvidia Tesla P100 GPU with a batch size of 32 and an average training time of 44 s/epoch. After hyper-parameter search, we selected the number of spatio-temporal blocks $K = 3$, the embedding dimension of the node dictionaries $d = 10$, the number of filters $f = 10$ and the kernel size $ks = 3$ for the temporal convolution layers across the model. The same configuration was selected for both tasks. Parameter sharing and the utilization of 1×1 convolutions across the model enables the development of an efficient model with 11205 trainable parameters. Our implementation is available open-source.[1]

3.3 Experimental Results

We compared the performance of DAST-GCN against the following baseline models:

- **SVM-linear**: a support vector machine with a linear kernel on the flattened Pearson's correlation matrix.

---

[1] https://github.com/AhmedElGazzar0/DAST-GCN.

– **SVM-rbf**: a support vector machine with a radial basis function kernel on the flattened Pearson's correlation matrix.
– **FCN**: a stack of fully connected layers on the flattened Pearson's correlation matrix.
– **BrainNetCNN** [10]: a 2D CNN on Pearson correlation matrix.
– **1D-CNN** [6]: a 1D CNN on the ROIs timecourses.
– **ST-GCN** [7]: a static ST-GCN where the nodes features represent ROI timecourses and the edges are defined using thresholded Pearson's correlation.

For a fair comparison, we conducted a hyper-parameter search for the all the baselines models with the exception of **BrainNetCNN** [10] and **ST-GCN** [7] where we used the default parameters recommended by the authors. Moreover, to evaluate the efficiency of our architecture design, we created ablated versions of the model; *DAST-GCN_tlc:* without temporal lag correction module, *DAST-GCN_M=1:* only one adjacency matrix is trained throughout the model, *DAST-GCN_corr:* adjacency matrix is pre-computed as the mean Pearson's correlation matrix of the training samples, *DAST-GCN_undir:* a symmetric undirected adjacency matrix, i.e. $E_t = E_s^T$. A comparison of the results of the model against baselines and ablated versions are shown in Table 1.

The results show superior performance of DAST-GCN against baseline models, more predominantly against correlation-based methods which suggest the importance of spatio-temporal modelling of brain activity. This is also supported by the performance of *ST-GCN*, *DAST-GCN_corr* and the results from [2] which proves the shortcomings of static pre-defined graph structure based on correlations. The ablation results verify the significance of accounting for the low temporal resolution of the signal before temporal feature extraction. Finally, the comparison of utilizing undirected or single adjacency matrix against $M = K$ dense graphs shows comparable results on sex classification and slightly more inferior results on age classification. Further, to assess the practical applicability and relevance of the proposed model in a clinical setting where the number of labeled samples is often scarce, we conducted the classification experiments for the two tasks on different scales of available data points per class (250, 500, 1000, 2500) and report the 5-fold balanced test accuracy results in Fig. 2.

4   Generalizability

A main challenge of applying deep learning models on neuroimaging datasets is the limited availability of labelled samples and the heterogeneity between fMRI datasets in terms of scanning parameters and participant demographics which restricts the application of pre-trained models across datasets. This explains the wide adoption of linear and shallow non-linear models on engineered features in neuroimaging. However, we hypothesize that a robust graph structure inferred for one task should be the invariant to different scanning parameters and demographics and thus DAST-GCN could be utilized to boost the performance on independent datasets with a limited number of labelled samples. We validated our hypothesis by transferring the graph structure trained on the UKBiobank

Table 1. 5-fold test results (mean ± standard deviation) of DASTG-GCN against baselines and ablated versions of the model.

Model	Sex			Age		
	Acc.%	Sens.%	Spec. %	Acc.%	Sens. %	Spec. %
SVM-linear	70.1 ± 1.9	69.9 ± 2.8	74.1 ± 3.4	54.3 ± 1.3	47.7 ± 1.4	55.7 ± 1.6
SVM-rbf	75.1 ± 1.7	77.4 ± 2.7	72.5 ± 4	57.8 ± 1.2	50.1 ± 1.3	66.8 ± 7
FCN	79.2 ± 1.0	79.5 ± 0.7	80.4 ± 2.5	63.1 ± 1.1	61.3 ± 1.5	61.1 ± 4.2
1D-CNN [6]	81.7 ± 1.4	82.7 ± 2.1	81.4 ± 0.8	65.9 ± 0.9	64.2 ± 1.1	66.4 ± 2
BrainNetCNN [10]	77.8 ± 1.8	78.1 ± 2.3	77.9 ± 2.9	60.9 ± 1.2	60.3 ± 1.8	62.5 ± 1.3
ST-GCN [7]	78.4 ± 2.3	77.2 ± 1.9	79.8 ± 1.8	60.5 ± 2.4	61.4 ± 1.9	60.1 ± 2.2
DAST-GCN_corr	78.7 ± 2.1	77.5 ± 1.8	79.3 ± 2.5	61.6 ± 2.9	61.1 ± 3.4	62.7 ± 5.6
DAST-GCN_tlc	82.6 ± 1.3	82.3 ± 1.2	82.1 ± 1.3	61.4 ± 2.6	60.6 ± 1.7	61.1 ± 1.3
DAST-GCN_undir	82.9 ± 1.4	83.1 ± 1.7	82.9 ± 1.4	64.7 ± 4.3	61.3 ± 2.4	63.1 ± 2.1
DAST-GCN_M=1	84.8 ± 0.7	84.9 ± 0.8	84.8 ± 0.7	64.3 ± 3.6	65.4 ± 2.7	64.3 ± 3.6
DAST-GCN	85.3 ± 0.6	85.4 ± 0.7	85.3 ± 0.7	68.6 ± 1.6	67.9 ± 1.5	68.4 ± 1.6

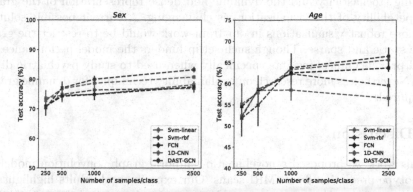

Fig. 2. 5-fold scaling performance for DAST-GCN at sex and age classifications against baseline models under different numbers of samples per class. Also shown are the standard-deviation across folds as error-bars.

dataset for sex classification to 120 healthy participants of the REST-meta-MDD dataset [23] (60 males/60 females, Age = 40.3 ± 15.6 years, TR = 1.3 s, T = 220 timepoints) for the same task. We chose the *DA-STGCN_M = 1* version of the model due to the different temporal resolution between datasets and lower number of parameters. We compared the performance of the pre-trained model (DA-STGCN_pretrained) against a model trained from scratch (DAST-GCN_no-pretrain), SVM-linear and SVM-rbf baselines. Test results of the 5-fold cross validation are reported in Table 2. Accuracy of the pretrained model was more than 7% point higher than SVM or without pretraining, with less than half the standard deviation. Our results support that the inferred graph structure is robust across datasets with different scanning parameters and demographics, and highlight the potential advantage of applying DAST-GCN in heterogeneous environments where datasets are collected from different sites.

Table 2. 5-fold test results (mean ± standard deviation) for sex classification on Rest-meta-MDD dataset.

Model	Acc.%	Sens.%	Spec.%
Svm-rbf	61.5 ± 6.1	65.9 ± 8.6	56.6 ± 20.4
Svm-linear	67.6 ± 10.2	70.7 ± 13.2	66.3 ± 12.4
DAST-GCN_nopretrain	66.6 ± 12.1	69.1 ± 12.6	62.4 ± 11.6
DAST-GCN_pretrained	**75.0 ± 4.56**	**79.6 ± 8.6**	**70.0 ± 10.1**

5 Limitations

A good fMRI learning model should not only provide good test metrics but also good explainability in order to advance clinical research. DAST_GCN infers the underlying graph structure while learning, hence offers a direct visualization of potential connectivity bio-markers for the task in hand. However, given training stochasticity and the dynamic and dense representation of the graph, interpetabillity of the visualizations is challenging. A current possible solution for more robust visualizations in our framework would be to restrict the graph to be static and sparse. Though such setup hinders the model performance, its could provide clinical insights specifically when used to study psychiatric disorders. Nevertheless more research on explainability techniques for dynamic graphs is required.

6 Discussion

In this work we proposed a novel spatio-temporal graph convolution model for phenotype prediction from fMRI scans. Our experimental results highlight the advantage of learning a dynamic graph structure versus pre-defined correlation based values. Further, we show that the task-inferred graph structure is robust against the heterogeneity of fMRI datasets and can be effectively transferred across different population demographics and scanning parameters. Finally, our results showcase the quantitative and qualitative advantage of applying deep learning models for phenotype prediction in neuroimaging at different scales and advocate for the development of models that incorporate proper inductive bias and operate on minimally pre-processed derivatives of the raw data.

Acknowledgement. This work was supported by the Netherlands Organization for Scientific Research (NWO; 628.011.023), Philips Research, AAA Data Science Program, and ZonMW (Vidi; 016.156.318).

References

1. Allen, E.A., Damaraju, E., Plis, S.M., Erhardt, E.B., Eichele, T., Calhoun, V.D.: Tracking whole-brain connectivity dynamics in the resting state. Cereb. Cortex **24**(3), 663–676 (2014)

2. Azevedo, T., Passamonti, L., Lio, P., Toschi, N.: Towards a predictive spatio-temporal representation of brain data. arXiv preprint arXiv:2003.03290 (2020)
3. Bai, L., Yao, L., Li, C., Wang, X., Wang, C.: Adaptive graph convolutional recurrent network for traffic forecasting. arXiv preprint arXiv:2007.02842 (2020)
4. Bruna, J., Zaremba, W., Szlam, A., LeCun, Y.: Spectral networks and locally connected networks on graphs. arXiv preprint arXiv:1312.6203 (2013)
5. Defferrard, M., Bresson, X., Vandergheynst, P.: Convolutional neural networks on graphs with fast localized spectral filtering. arXiv preprint arXiv:1606.09375 (2016)
6. El Gazzar, A., Cerliani, L., van Wingen, G., Thomas, R.M.: Simple 1-D convolutional networks for resting-state fMRI based classification in autism. In: 2019 International Joint Conference on Neural Networks (IJCNN), pp. 1–6. IEEE (2019)
7. Gadgil, S., Zhao, Q., Pfefferbaum, A., Sullivan, E.V., Adeli, E., Pohl, K.M.: Spatio-temporal graph convolution for resting-state fMRI analysis. In: Martel A.L., et al. (eds.) MICCAI 2020. LNCS, vol. 12267, pp. 528–538. Springer, Cham (2020). https://doi.org/10.1007/978-3-030-59728-3_52
8. Gilmer, J., Schoenholz, S.S., Riley, P.F., Vinyals, O., Dahl, G.E.: Neural message passing for quantum chemistry. In: International Conference on Machine Learning, pp. 1263–1272. PMLR (2017)
9. He, T., et al.: Deep neural networks and kernel regression achieve comparable accuracies for functional connectivity prediction of behavior and demographics. NeuroImage 206, 116276 (2020)
10. Kawahara, J., et al.: BrainNetCNN: convolutional neural networks for brain networks; towards predicting neurodevelopment. NeuroImage 146, 1038–1049 (2017)
11. Kipf, T.N., Welling, M.: Semi-supervised classification with graph convolutional networks. arXiv preprint arXiv:1609.02907 (2016)
12. Mitra, A., Snyder, A.Z., Hacker, C.D., Raichle, M.E.: Lag structure in resting-state fMRI. J. Neurophysiol. 111(11), 2374–2391 (2014)
13. van den Oord, A., et al.: Wavenet: a generative model for raw audio. arXiv preprint arXiv:1609.03499 (2016)
14. van den Oord, A., Kalchbrenner, N., Vinyals, O., Espeholt, L., Graves, A., Kavukcuoglu, K.: Conditional image generation with pixelCNN decoders. arXiv preprint arXiv:1606.05328 (2016)
15. Rubinov, M., Sporns, O.: Complex network measures of brain connectivity: uses and interpretations. Neuroimage 52(3), 1059–1069 (2010)
16. Schulz, M.A., et al.: Different scaling of linear models and deep learning in UKBiobank brain images versus machine-learning datasets. Nat. Commun. 11(1), 1–15 (2020)
17. Sporns, O., Tononi, G., Kötter, R.: The human connectome: a structural description of the human brain. PLoS Comput. Biol. 1(4), e42 (2005)
18. Tzourio-Mazoyer, N., et al.: Automated anatomical labeling of activations in SPM using a macroscopic anatomical parcellation of the MNI MRI single-subject brain. Neuroimage 15(1), 273–289 (2002)
19. Vieira, S., Pinaya, W.H., Mechelli, A.: Using deep learning to investigate the neuroimaging correlates of psychiatric and neurological disorders: methods and applications. Neurosci. Biobehav. Rev. 74, 58–75 (2017)
20. Wang, L., Li, K., Hu, X.P.: Graph convolutional network for fmri analysis based on connectivity neighborhood. Network Neuroscience pp. 1–13
21. Wu, Z., Pan, S., Chen, F., Long, G., Zhang, C., Philip, S.Y.: A comprehensive survey on graph neural networks. IEEE Trans. Neural Netw. Learn. Syst. 32, 4–24 (2020)

22. Wu, Z., Pan, S., Long, G., Jiang, J., Zhang, C.: Graph wavenet for deep spatial-temporal graph modeling. arXiv preprint arXiv:1906.00121 (2019)
23. Yan, C.G., et al.: Reduced default mode network functional connectivity in patients with recurrent major depressive disorder. Proc. Natl. Acad. Sci. **116**(18), 9078–9083 (2019)
24. Yan, S., Xiong, Y., Lin, D.: Spatial temporal graph convolutional networks for skeleton-based action recognition. In: Proceedings of the AAAI Conference on Artificial Intelligence, vol. 32 (2018)
25. Zhang, Y., Tetrel, L., Thirion, B., Bellec, P.: Functional annotation of human cognitive states using deep graph convolution. NeuroImage **231**, 117847 (2021)

Structure-Function Mapping via Graph Neural Networks

Yang Ji[⊠][iD], Samuel Deslauriers-Gauthier[iD], and Rachid Deriche[iD]

Université Côte d'Azur, Inria, France
{Yang.Ji,Samuel.Deslauriers-gauthier,Rachid.Deriche}@inria.fr

Abstract. Understanding the mapping between structural and functional brain connectivity is essential for understanding how cognitive processes emerge from their morphological substrates. Many studies have investigated the problem from an eigendecomposition viewpoint, however, few have taken a deep learning viewpoint, even less studies have been engaged within the framework of graph neural networks (GNNs). As deep learning has produced significant results in several fields, there has been an increasing interest in applying neural networks to graph problems. In this paper, we investigate the structural connectivity and functional connectivity mapping within a deep learning GNNs based framework, including graph convolutional networks (GCN) and graph transformer networks (GTN). To our knowledge, this original GTN based framework has never been studied in the context of structure-function and brain connectivity mapping. To achieve this goal, we use a GNNs based encoder-decoder system, where the encoder takes structural connectivity (SC) matrix as input and generates a latent representation of each node in a lower dimension, then the decoder uses the latent representation to reconstruct or predict the associated functional connectivity (FC) matrix. Besides comparing different encoders for node embedding, we also demonstrate that a decoder, which projects lower dimension vectors onto higher dimensional space, can improve the model performance. Our experiments demonstrate that both GCN encoder and GTN encoder combined with the proposed decoder can provide better results on our data than the previously proposed GCN autoencoder model. GTN encoder is also shown to be much more effective when it comes to noisy data and outliers.

Keywords: Brain connectivity mapping · Graph neural network · Graph convolutional network · Graph transformer network · Autoencoder

1 Introduction

Studying the relationship between structural connectivity and functional connectivity is essential to understand how brain function emerges from their underlying structural substrate. A better characterization of the link between SC and FC

© Springer Nature Switzerland AG 2021
A. Abdulkadir et al. (Eds.): MLCN 2021, LNCS 13001, pp. 135–144, 2021.
https://doi.org/10.1007/978-3-030-87586-2_14

will provide insights into how lesions to the structural substrate affect brain function [1–3], with great potential clinical applications. Previous studies have shown that functional connectivity and structural connectivity patterns are correlated [4–6]. Therefore, it is reasonable to expect that FC matrices can be predicted from SC matrices.

Many models have been proposed to map structure to function [9–13]. A subset of these models are specifically based on the link between the eigenvalues and eigenvectors of SC and FC matrices. However, this approach is based on the assumption that there exists a correspondence between structural and functional eigenmodes that can be captured by the models. Moreover, they are usually computationally expensive when dealing with large data. In contrast, deep learning (DL) approaches provide a more general approach to learn the mapping between SC and FC. Deep learning has produced significant results in several fields, including computer vision, speech recognition and natural language processing [7]. To integrate non-Euclidean graph data, graph neural networks (GNNs) have been developed to deal with graph problems [8]. For example, graph convolutional networks (GCN), which were proposed by [14], are motivated by the graph convolution operator [15]. At a higher level, GCNs use the graph Fourier transform to aggregate neighboring nodes' features, thus exploiting the topology of the graph during the learning process. GCNs have been studied on structure-function mapping in [16], where a graph encoder-decoder based on GCNs was proposed to recover the SC-FC mapping. However, the decoder in this model is mainly an inner product operation with no trainable parameters. The linear encoder used by the model can also be sensitive to outliers presented in data. A second example are graph transformer networks (GTN) [17] are analogous to Spatial Transformer Networks [18]. The most advantageous point of GTNs over GCNs is that they consider heterogeneous graph structures, and can therefore be used to predict FC from several different SC matrices. This feature makes GTN more robust when dealing with outliers or noisy data, where GCN is much less effective.

In this work, we revisit structure-function mapping via deep graph neural networks. First, we investigate the importance of a trainable decoder in the GCN architecture of [16] and show that it improves the prediction of FC from SC. Second, we propose a GTN architecture to predict FC from several diffusion derived SC matrices and show that is more robust to outliers. To our knowledge, this is the first time GTNs have been studied in the context of structure-function and brain connectivity mapping. All of the models are tested on 1050 subjects of the Human Connectome Project, therefore providing reproducible results.

The rest of the paper is composed of three parts. The first part is Preliminaries, where we state the main problem and then introduce the mathematical theories of GCN, GTN and encoder-decoder system. Then we propose numerous models based on GNNs. The second part is Experiments, where we present how SC/FC matrices are extracted from Magnetic resonance imaging (MRI), and details of the implementation. The final part is Results and conclusion, where we compare and discuss in detail the performance of each proposed model.

2   Preliminaries

In this section, we first state the problem and the objective of this paper and review some GNNs architectures including GCN, GTN, and the main framework Autoencoder that includes them all. At the end of each architecture description, a model is proposed.

2.1   Problem Statement

Given an undirected SC matrix, which is extracted from diffusion MRI, we define $A \in \mathbb{R}^{N \times N}$ as its symmetric adjacency matrix, $D_{ii} = \sum_{j=0}^{N} A_{ij}$ as the diagonal degree matrix which denotes the number of connections of each node, N as the number of nodes and in our case it is the number of regions of the atlas. The normalized Laplacian of A is defined as $L = I_N - D^{-\frac{1}{2}} A D^{-\frac{1}{2}} = U \Lambda U^T$, where Λ denotes its eigenvalues and U as its eigenvector. The brain connectivity mapping is between the SC matrix and the FC matrix. The objective is to identify a mapping that takes A as input and predicts a FC matrix $Z \in \mathbb{R}^{N \times N}$, such that Z is close to the empirical FC matrix. The problem can be reformulated as

$$\text{minimize } \frac{1}{K} \sum_{k=1}^{K} \|f(A_k) - Z_k\|_F^2 \tag{1}$$

where f is the mapping function (in our case a DL architecture), K is the number of subjects, and $\| \cdot \|_F$ is the Frobenius norm.

2.2   Autoencoder

The idea of autoencoder [19] is to use an encoder to learn a latent representation of the data, then use a decoder to reconstruct its input. Here, we use the autoencoder as our main structure, but we alter it into reconstructing the associated FC matrix instead of SC matrix. An encoder generates a latent vector representation y_i for each node, then a decoder uses these representation vectors to compute the prediction of the FC matrix, denoted as $Z' = f'(Y)$ with $Y \in \mathbb{R}^{N \times F}$ where F is the dimension of the latent node representation. In this work, we test both GCN (as in [16]) and GTN as encoders. A diagram of the model is presented in Fig. 1.

The decoder mainly serves to transform latent nodes embedding Y to target symmetric FC prediction Z. In [16], the authors proposed a decoder based on inner product, defined as:

$$Z' = \tanh(\text{ReLU}(YY^T)) \tag{2}$$

where $\tanh(x) = \frac{e^x - e^{-x}}{e^x + e^{-x}}$ and $\text{ReLU}(x) = max(0, x)$ are activation functions. Z' is a symmetric matrix whose values are in the range $[0, 1]$, like the observed FC matrix. However, apart from reconstructing a prediction that fits the FC matrix'

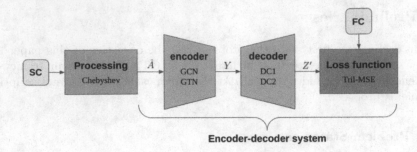

Fig. 1. Model pipeline. The first step is to process SC matrix to obtain $\hat{A}$. Then the encoder (GCN or GTN) takes $\hat{A}$ as input and generate embedding Y. Finally, the decoder (DC1 or DC2) takes Y as input and reconstruct FC matrix Z'. A loss function (Tril-MSE) calculates the error between Z' and the empirical FC.

symmetry and range property, it is not trainable. In the rest of the article, this decoder is referred to as **DC1**.

To overcome the disadvantage of DC1, we propose a decoder that is trainable, defined as

$$Y' = YW_0^d + B^d,$$
$$Y'' = \tanh(Y' * W_1^d),$$
$$Z' = \mathrm{ReLU}(Y''Y''^T) \tag{3}$$

where $W_0^d \in \mathbb{R}^{F \times N}$ is a trainable parameter which augments the dimension of the latent embedding Y, $B^d \in \mathbb{R}^{N \times N}$ is a bias parameter, $W_1^d \in \mathbb{R}^{N \times 1}$ is also a trainable parameter, initialized with constant 1 to keep the output's range close to $[0, 1]$ from the start. It balances the dynamic of the inner product operation. The $*$ operator denotes a row-wise multiplication. To predict a semidefinite positive FC matrices, for example when FC matrices are not thresholded, it suffices to omit $\mathrm{ReLU}(\cdot)$. In the rest of the article, this decoder is referred to as **DC2**.

2.3 Graph Convolutional Networks (GCN)

A spectral graph convolution operator, as defined by [22], is given by

$$g_\theta \star x = U g_\theta U^T x \tag{4}$$

with the filter $g_\theta = \mathrm{diag}(\theta)$, $\theta \in \mathbb{R}^N$ is the parameter. Therefore g_θ can be considered as a function of the eigenvalues of the Laplacian Λ. In order to avoid expensive computation of eigendecomposition of graph Laplacian L, [20] proposed using Chebyshev polynomials to approximate $g_\theta(\Lambda)$. Given the recursively defined Chebyshev polynomial as $T_k(x) = 2xT_{k-1}(x) - T_{k-2}(x)$, with $T_0 = 1, T_1 = x$, $g_\theta(\Lambda)$ is reformulated as

$$g_{\theta'}(\Lambda) \approx \sum_{k=0}^{K} \theta'_k T_k(\tilde{\Lambda}) \tag{5}$$

where $\tilde{\Lambda} = \frac{2}{\lambda_{max}}\Lambda - I_N$, λ_{max} is the largest eigenvalue of L. $\theta^{'} \in \mathbb{R}^K$ is the coefficient vector of Chebyshev polynomial. Combining Eqs. (4) and (5), a new graph convolution operator can be defined as

$$g_{\theta'} \star x \approx \sum_{k=0}^{K} \theta_k' T_k(\tilde{L})x \tag{6}$$

where $\tilde{L} = \frac{2}{\lambda_{max}}L - I_N$.

In [14], a simple graph convolution operator is proposed, which is motivated by a first-order approximation of Chebyshev polynomials [20]. By limiting $K = 1$ and setting $\lambda_{max} = 2, \theta_0' = -\theta_1' = \theta$ in Eq. (6), we have the graph convolution operator that is motivated by the first order of Chebyshev

$$g_{\theta'} \star x \approx \theta(I_N + D^{-\frac{1}{2}}AD^{-\frac{1}{2}})x = \theta\tilde{D}^{-\frac{1}{2}}\tilde{A}\tilde{D}^{-\frac{1}{2}}x \tag{7}$$

where $\tilde{A} = A + I_N$ is a symmetric self-looped adjacency matrix, $\tilde{D}_{ii} = \sum_{j=0}^{N}\tilde{A}_{ij}$ denotes the degree matrix of the $\tilde{A}$. Finally the first order approximation GCN is defined as

$$Y^{(l+1)} = \sigma\left(\tilde{D}^{-\frac{1}{2}}\tilde{A}\tilde{D}^{-\frac{1}{2}}Y^{(l)}W_0^{(l)}\right), \tag{8}$$

where $Y^0 = X$, $Y^{(l)}$ denotes the lth hidden layer, $W_0^{(l)}$ is the weight parameter of lth layer. $\sigma(\cdot)$ is an activation function.

In our implementation, the GCN encoder is a single layer GCN defined as

$$Y = \text{ReLU}(\hat{A}XW_0), \tag{9}$$

where $\hat{A} = \tilde{D}^{-\frac{1}{2}}\tilde{A}\tilde{D}^{-\frac{1}{2}}$, $\tilde{A}$ was defined in Eq. (7), $W_0 \in \mathbb{R}^{N \times F}$ is a trainable parameter of the encoder, N and F are previously defined.

2.4 Graph Transformer Networks (GTN)

Graph Transformer Networks [17] take heterogeneous graphs as multi-channel input and use these channels to compute multi-channel meta-path graph tensors. Then, GTNs apply a single layer GCN to each channel of the meta-path tensor and generate a node representation for each channel. Finally, a perceptron layer is added on top of the concatenation of the node representations of each channel. Consider $\mathbb{A} \in \mathbb{R}^{N \times N \times C}$ a set of adjacency matrices of heterogeneous graphs, C denotes channel number, which in our case $\mathbb{A} = [\hat{A}_1, \hat{A}_2, \hat{A}_3]$ is composed by the first order approximation of three types of SC matrices, whose nature is detailed in Sect. 3.1. First the Graph Transformer (GT) layer generates two intermediate adjacency matrices, $\mathbb{Q}_1 \in \mathbb{R}^{N \times N \times C}$ and $\mathbb{Q}_2 \in \mathbb{R}^{N \times N \times C}$, then performs a 1×1 convolution as

$$\mathbb{Q}_i = \phi(\mathbb{A}, \text{softmax}(W^{\phi})), \tag{10}$$

where ϕ is a graph convolution, $W^{\phi} \in \mathbb{R}^{1 \times 1 \times K}$ is the parameter of ϕ, softmax($\cdot$) is an activation function defined as softmax(x_i) = $\frac{exp(x_i)}{\sum_j exp(x_j)}$. Second, the GT layer combines the intermediate adjacency matrices Q_i together to compute a normalized new graph structure. For the $A^{(1)} = \mathbb{D}_{(1)}^{-1} Q_1 Q_2$, $\mathbb{D} \in \mathbb{R}^{N \times N \times C}$ is the multi channels degree tensor of $Q_1 Q_2$. For multi layers GTN, we repeat the procedure from Eq. (10) by generating Q_{l+1}, then we compute the new graph $A^{(l)} = \mathbb{D}_{(l)}^{-1} A^{(l-1)} Q^{(l+1)}$. Finally, a graph convolution is performed on each channel of the new structure $A^{(l)}$ before combining all channels' node embedding. Instead of concatenating all channels' embedding as in the original paper [17], we found that summing them performs better. Thus the intermediate output is

$$Y' = \sum_{i=1}^{C} \sigma \left(\tilde{D}_i^{-1} \tilde{A}_i^{(1)} X W_0 \right) \tag{11}$$

where $\tilde{A}_i^{(l)} = A_i^{(l)} + I$. C denotes the number of channels, $W_0 \in \mathbb{R}^{N \times N}$ is a trainable weight parameter. In our implementation of GTN, we set the layer number to $l = 1$ and the channel number to $C = 3$. Upon the output defined in Eq. (11) in Sect. 2.4, an additional layer is added in order to obtain lower dimension embedding. The ultimate GTN encoder is defined as

$$Y = \text{ReLU} \left(Y' W_1 + B_1 \right) \tag{12}$$

where $W_1 \in \mathbb{R}^{N \times F}$ is a trainable parameter, $B_1 \in \mathbb{R}^{F \times F}$ denotes bias.

3 Experiments

3.1 Data

The SC and FC dataset, which contain 1050 subjects, is provided by Human Connectome Project[1] (HCP). The SC and FC data are extracted from diffusion MRI (dMRI) and functional MRI (fMRI) using the same pipelines as in [28]. We used the streamlines produced by the diffusion pipeline to compute three types of SC matrices, $Count, SIFT2, Length$. The first one $Count(i,j)$ represents the total number of streamlines between regions i and j. The second one $SIFT2(i,j)$ was computed by summing the SIFT2 weights of streamlines connecting i and j [29]. The third one $Length(i,j)$, denotes the reciprocal average length of the streamlines connecting i and j. Both SC and FC adjacency matrix contain $N = 68$ rows and columns, each row and column corresponds to a brain region of the Desikan-Killiany atlas.

The SC matrix A extracted from dMRI is not directly used as model input. First, we set the diagonal values of A to be zero. Then, as explained in Sect. 2.3, we use first order approximation of Chebyshev polynomial on A to obtain the input $\hat{A}$, defined in Eq. (7). As $Length$ data has infinite values (between disconnected brain regions), for computational stability, we set its diagonal values as 1, then compute $1/Length(i,j)$ beforehand. The FC originally ranges between

---

[1] http://www.humanconnectomeproject.org/.

−1 and 1, but in our implementation, the negative values in FC are set to 0 as in previous researches dealing with functional brain connectivity [26,27]. After our processing pipeline, the average correlation between the FC matrix and the *Count*, *Length*, and *SIFT2* matrices is 0.24, 0.28, and 0.24, respectively.

3.2 Implementation

The GCN uses only the *Count* matrix, while GTN uses all three SC matrices mentioned previously. The dataset is split into three parts: 80% for training, 10% for validation and 10% for test. Cross-validation (10 folds) and early stop are applied. We use Adam as optimizer [23], set learning rate to 0.001; batch to 25 for GCN, 1 for GTN; dropout rate to 0.05; encoder's output dimension to $F = 32$ for all models. The weight parameters are initialized using [24]. The models are implemented via Pytorch [25]. We use Mean Square Error (MSE) as all models' loss function, however, it is only applied on the lower triangular part of the predicted FC matrix, since FC matrix is symmetric and whose diagonal values are always 1. We reference it as **Tril-MSE** in the following. We also use Pearson Correlation as a supplementary measure to double evaluate the quality of predictions, although we do not train our models on it. Same as Tril-MSE, we apply it only on the lower triangular part of output matrices, we reference it as **TPearson** for the rest of the paper.

4 Results and Discussion

We evaluate the average and median Tril-MSE test error, as well as the average and median TPearson between predicted FC and empirical FC. It is also important to evaluate the median test error because outliers are often present in data and difficult to identify beforehand. A model with a lower median test error is more robust when it comes to a dataset with outliers, an important consideration in clinical applications.

As presented in Table 1, the first architecture "GCN+DC1" from [16] is the baseline for comparison, while the rest are the new models proposed in this work. We observe that both GCN and GTN encoder combined with DC2, defined in Eq. (3) have obtained better scores in all measures than when they are combined with DC1, defined in Eq. (2). This improvement can be explained by the additional trainable parameters of the decoder which projects the latent space to the FC prediction. When comparing the median Tril-MSE, we can notice that GTN increases the performance by up to 18% when compared to GCN. This significant change in the median error highlights the robustness of GTN to outliers. In addition, GTN also outperforms GCN at the supplementary measure TPearson. Also note that the GTN always provides a better FC prediction regardless of the decoder selected, showing that important information is captured in the *Length*, and *SIFT2* matrices not exploited by the GCN architecture.

Table 1. Average and median test error for the graph convolutional and transformer networks using different decoders: DC1 and DC2 are defined in Eq. (2) and (3), respectively.

Results	Average Tril-MSE	Median Tril-MSE	Average TPearson	Median TPearson
GCN+DC1	0.0430 ± 0.0016	0.0431	0.6628 ± 0.0077	0.6671
GCN+DC2	0.0404 ± 0.0018	0.0402	0.7003 ± 0.0073	0.7069
GTN+DC1	0.0401 ± 0.0020	0.0351	0.7160 ± 0.0077	0.7228
GTN+DC2	$\mathbf{0.0397 \pm 0.0018}$	**0.033**	$\mathbf{0.7154 \pm 0.0061}$	**0.7231**

5 Conclusion

A better understanding of the link between SC and FC has the potential to provide insights into how lesions to the structural substrate affect brain function. In this work, we have revisited the structure-function mapping via graph neural networks. We investigated the importance of a trainable decoder in a previously proposed GCN architecture and proposed a novel GTN architecture to predict FC from several diffusion derived SC matrices. We presented prediction results on 1050 subjects of the Human Connectome Project, showing that trainable parameters in the decoder improve the performance of the models. In addition, our results indicated that GTNs, which simultaneously exploit different realisations of the SC matrix, outperformed GCNs. Overall, the preliminary results presented in this work highlight the potential for graph neural networks in structure-function mapping.

Acknowledgment. This work has been supported by the ERC under the European Union's Horizon 2020 research and innovation program (ERC Advanced Grant agreement No 694665: CoBCoM: Computational Brain Connectivity Mapping) and by the French government, through the 3IA Côte d'Azur Investments in the Future project managed by the National Research Agency (ANR) with the reference number ANR-19-P3IA-0002.

Data were provided by the Human Connectome Project, WU-Minn Consortium (Principal Investigators: David Van Essen and Kamil Ugurbil; 1U54MH091657) funded by the 16 NIH Institutes and Centers that support the NIH Blueprint for Neuroscience Research; and by the McDonnell Center for Systems Neuroscience at Washington University.

The authors are grateful to Inria Sophia Antipolis - Méditerranée https://wiki.inria.fr/ClustersSophia/Usage_policy "Nef" computation cluster for providing resources and support.

The authors are grateful to the OPAL infrastructure from Université Côte d'Azur for providing resources and support.

References

1. Sporns, O., Tonomi, G., Kötter, R.: The human connectome: a structural description of the human brain. PLoS Comput. Biol. **1**, 245–251 (2005)
2. Alstott, J., Breakspear, M., Hagmann, P., Cammoun, L., Sporns, O.: Modeling the impact of lesions in the human brain. PLoS Comput. Biol. **5**. Article ID e1000408
3. Váša, F., Shanahan, M., Hellyer, P., Scott, G., Cabral, J., Leech, R.: Effects of lesions on synchrony and metastabistability in cortical networks. NeuroImage **118**, 456–467
4. Honey, C., et al.: Predicting human resting-state functional connectivity from structural connectivity. Proc. Natl. Acad. Sci. U.S.A. **106**(6), 2035–2040 (2009)
5. Deco, G., Kringelbach, M.L., Jirsa, V.K., Ritter, P.: The dynamics of resting fluctuations in the brain: metastability and its dynamical cortical core. Sci. Rep. 7, 3095, 2
6. Hermundstad, A.M., Bassett, D.S., Brown, K.S., et al.: Structural foundations of resting-state and task-based functional connectivity in the human brain. Proc. Nat. Acad. Sci. **110**(15), 6169–6174
7. LeCun, Y., Bengio, Y., Hinton, G.: Deep learning. Nature **521**(7553), 436 (2015)
8. Wu, Z., Pan, S., Chen, F., Long, G., Zhang, C., Yu, P.S.: A comprehensive survey on graph neural networks. IEEE Trans. Neural Netw. Learn. Syst. **32**(1), 4–24 (2021). https://doi.org/10.1109/TNNLS.2020.2978386
9. Deligianni, F., et al.: A framework for inter-subject prediction of functional connectivity from structural networks. IEEE Trans. Med. Imaging **32**(12), 2200–2214 (2013)
10. Abdelnour, F., Voss, H.U., Raj, A.: Network diffusion accurately models the relationship between structural and functional brain connectivity networks. Neuroimage **90**, 335–347 (2014)
11. Meier, J., et al.: A mapping between structural and functional brain networks. Brain Connect. **6**(4), 298–311 (2016)
12. Liang, H., Wang, H.: Structure-function network mapping and its assessment via persistent homology. PLoS Comput. Biol. **13**(1), e1005325 (2017)
13. Becker, C.O., et al.: Spectral mapping of brain functional connectivity from diffusion imaging. Sci. Rep. **8**(1) 1411 (2018)
14. Kipf, T.N., Welling, M.: Semi-supervised classification with graph convolutional networks. In: ICLR (2017)
15. Bruna, J., Zaremba, W., Szlam, A., LeCun, Y.: Spectral networks and locally connected networks on graphs. In: Proceedings of ICLR (2014)
16. Li, Y., Shafipour, R., Mateos, G., Zhang, Z.: Mapping brain structural connectivities to functional networks via graph encoder-decoder with interpretable latent embeddings. In: IEEE Global Conference on Signal and Information Processing (GlobalSIP) 2019, pp. 1–5 (2019). https://doi.org/10.1109/GlobalSIP45357.2019.8969239
17. Yun, S., Jeong, M., Kim, R., Kang, J.: Graph transformer networks. In: NeurIPS (2019)
18. Jaderberg, M., Simonyan, K., Zisserman, A., et al.: Spatial transformer networks. In: Advances in Neural Information Processing Systems, pp. 2017–2025 (2015)
19. Rumelhart, D.E., Hinton, G.E., Williams, R.J.: Parallel distributed processing: explorations in the microstructure of cognition. In: Learning Internal Representations by Error Propagation, vol. 1, pp. 318–362. MIT Press, Cambridge (1986)

20. Hammond, D.K., Vandergheynst, P., Gribonval, R.: Wavelets on graphs via spectral graph theory. Appl. Comput. Harmon. Anal. **30**(2), 129–150 (2011)
21. Venkatesh, M., Jaja, J., Pessoa, L.: Comparing functional connectivity matrices: a geometry-aware approach applied to participant identification. NeuroImage **207**, 116398 (2020). https://doi.org/10.1016/j.neuroimage.2019.116398. ISSN 1053-8119
22. Defferrard, M., Bresson, X., Vandergheynst, P.: Convolutional neural networks on graphs with fast localized spectral filtering. In: Advances in Neural Information Processing Systems, pp. 3844–3852 (2016)
23. Kingma, D.P., Ba, J.: Adam: a method for stochastic optimization. In: ICLR (2015)
24. Glorot, X., Bengio, Y.: Understanding the difficulty of training deep feedforward neural networks. In: International Conference on Artificial Intelligence and Statistics, pp. 249–256 (2010)
25. Paszke, A., et al.: Automatic Differentiation in PyTorch. In: NIPS 2017 Workshop on Autodiff (2017)
26. Power, J.D., Fair, D.A., Schlaggar, B.L., Petersen, S.E.: The development of human functional brain networks. Neuron **67**(5), 735–748 (2010)
27. Rubinov, M., Sporns, O.: Complex network measures of brain connectivity: uses and interpretations. Neuroimage **52**(3), 1059–1069 (2010)
28. Deslauriers-Gauthier, S., Zucchelli, M., Frigo, M., Deriche, R.: A unified framework for multimodal structure-function mapping based on eigenmodes. Med. Image Anal, p. 22 (2020). https://doi.org/10.1016/j.media.2020.101799.hal-02925913
29. Smith, R.E., Tournier, J.D., Calamante, F., Connelly, A.: SIFT2: enabling dense quantitative assessment of brain white matter connectivity using streamlines tractography. Neuroimage **119**, 338–351 (2015). https://doi.org/10.1016/j.neuroimage.2015.06.092. PMID: 26163802

Improving Phenotype Prediction Using Long-Range Spatio-Temporal Dynamics of Functional Connectivity

Simon Dahan[1(✉)], Logan Z. J. Williams[1,2], Daniel Rueckert[3,4], and Emma C. Robinson[1,2]

[1] Department of Biomedical Engineering, School of Biomedical Engineering and Imaging Sciences, King's College London, London SE1 7EH, UK
`simon.dahan@kcl.ac.uk`
[2] Centre for the Developing Brain, Department of Perinatal Imaging and Health, School of Biomedical Engineering and Imaging Sciences, King's College London, London SE1 7EH, UK
[3] Biomedical Image Analysis Group, Department of Computing, Imperial College London, London SW7 2AZ, UK
[4] Klinikum rechts der Isar, Technical University of Munich, 81675 Munich, Germany

Abstract. The study of functional brain connectivity (FC) is important for understanding the underlying mechanisms of many psychiatric disorders. Many recent analyses adopt graph convolutional networks, to study non-linear interactions between functionally-correlated states. However, although patterns of brain activation are known to be hierarchically organised in both *space and time*, many methods have failed to extract powerful spatio-temporal features. To overcome those challenges, and improve understanding of long-range functional dynamics, we translate an approach, from the domain of skeleton-based action recognition, designed to model interactions across space and time. We evaluate this approach using the Human Connectome Project (HCP) dataset on sex classification and fluid intelligence prediction. To account for subject topographic variability of functional organisation, we modelled functional connectomes using multi-resolution dual-regressed (subject-specific) ICA nodes. Results show a prediction accuracy of 94.4% for sex classification (an increase of 6.2% compared to other methods), and an improvement of correlation with fluid intelligence of 0.325 vs 0.144, relative to a baseline model that encodes space and time separately. Results suggest that explicit encoding of spatio-temporal dynamics of brain functional activity may improve the precision with which behavioural and cognitive phenotypes may be predicted in the future.

Keywords: Human connectome project · Graph convolution networks · Temporal dynamics · Functional MRI · Phenotyping

1 Introduction

Functional MRI (fMRI) is widely recognised as a cornerstone technique for relating brain processes to behaviour [15,24]. In particular, the correlation between

© Springer Nature Switzerland AG 2021
A. Abdulkadir et al. (Eds.): MLCN 2021, LNCS 13001, pp. 145–154, 2021.
https://doi.org/10.1007/978-3-030-87586-2_15

regional time-series activations, referred to as functional connectivity (FC), is known to be subject-specific [4,11] and discriminate between populations [13,23]. This has led to it being popularly investigated as a potential clinical biomarker for many psychiatric and neurological disorders [9,13]. Despite several promising studies on large cohorts, for example, the Autism Spectrum Disorder study ABIDE [13], several recent works have emphasised the need to better account for morphological, or topographic, heterogeneity of cortical architecture when building functional networks [2,6,12]. This is considered especially important for disease modelling [17], where phenotypes are often heterogeneous and may be classified into sub-types [28]. Under such circumstances fitting all data to a single global-average model of healthy functional organisation, stands to mask subtle features of disease [17,28], as well as obscure understanding of mechanisms of disorders that usually co-occur [28]. For these reasons, increasing efforts are being made to look towards individual subject-level analyses and move away from the case-control approach [18].

To this end, recent approaches have looked towards better capture of inter-subject variation when deriving functional networks. For example, probabilistic functional modes (or PROFUMO) [8] uses a variational form of independent component analysis (ICA) to explicitly account for subject-specific variability during group-wise factorisation of fMRI data into temporal and spatial modes; [11] implements a multi-session hierarchical Bayesian model (MS-HBM) in order to cluster functional connectivity data into group-wise comparable, but subject-specific, functional parcellations; and [6] projects a hand-annotated, population-average multimodal parcellation of the cortex onto individuals by training a multi-layer perceptron (MLP) classifier; in the process revealing considerable topographic variation of regions when observed across the group.

At the same time, many studies are moving away from modelling static functional connectivity (averaged over all time) towards explicit modelling of temporal dynamics, as a means of better capturing temporal markers of behaviour [5,10,19,27]. Of particular note, for this paper, are methods which translate models used for video understanding [29], especially graph convolution networks (GCN) used in skeleton-based action recognition [5]. In this paper we specifically investigate the potential for translation of new techniques which explicitly encode interactions, across nodes, in both space and time [16]. The specific contributions of this paper are as follows:

1. **Explicit modelling of FC spatio-temporal dynamics improves prediction of behavioural phenotypes**: We demonstrate improvements over existing GCN models of dynamic FC [5] by adopting a new model [16], which aggregates functional signals in both space and time.
2. **Accounting for inter-subject cortical heterogeneity improves performance**: We compare models trained using a 22 region group-average parcellation and multi-resolution dual-regressed (subject-specific) ICA nodes, and demonstrate gains in performance by accounting for topographic variability of functional organisation.

2   Related Works

The use of GCNs to study FC supports learning of complex, non-linear, long-range interactions between correlated functional network states. While early approaches focused on the analysis of static FC [3,13,14], demonstrating improved performance in metric learning for behavioural prediction of neuropsychiatric disorders such as Autism Spectrum Disorder [13,14]; more recent papers have started to investigate the potential of GCNs to model FC temporal dynamics [5,10]. Of these, [5] takes inspiration from a skeleton-based GCN model for video understanding [29], translating this to the domain of dynamic FC modelling, by building a spatial graph from static FC (estimated across all node time series) but self-connecting nodes temporally along a regular grid. By applying spatial-only and temporal-only convolutions, the method successfully captured long-range (spatial and temporal) dependencies but failed to extract powerful space-time features by decoupling space and time graphs. By contrast, the more recent MS-G3D of [16] goes further to capture spatio-temporal semantics between nodes by encoding them through novel graph convolutional units which aggregate across space and time; in this way encoding complex actions, for example, the action of sitting down or standing up, in which the top half of a skeleton moves before the bottom. In this paper, we hypothesise that such long-range spatio-temporal behaviour is also observed in brain functional dynamics. We, therefore, benchmark this method against the method proposed by [5] as a means for quantifying this effect.[1]

3   Methods

Data. This study used functional data from the first session (15 min - 1200 frames - 0.72 s/frame) of the Human Connectome Project (HCP) S1200 release [25], where individual subject node timeseries for 1003 participants were derived following group-wise independent component analysis (Group-ICA) [1,24]. In brief, individual functional MRI data was processed through the HCP pipelines [7,24] and aligned using MSMAll [21,22]. Then, spatial Group-ICA (with 468 subjects) was applied at different dimensionalities (15 to 300), factorising the data into a set of spatial maps (functional nodes) and associated time courses. These were propagated into individuals using dual regression, in this way tailoring the brain parcellations and time courses at the individual level (illustrated in Fig. 1). We used all available data for the sex classification task (534 males and 469 females), but had to exclude 4 subjects for the fluid intelligence prediction task (due to missing data). We used Penn Progressive Matrices as scores of fluid intelligence, referred as *PMAT24_A_CR* in the HCP behavioral data. Similarly to [5], we also benchmark against use of *group-average* functional parcellation, where the group average HCP multimodal cortical parcellation (180 regions per hemisphere) [6] was aggregated into 22 regions of interest (ROIs) per hemisphere

---

[1] The code for this experiment will be made available at: http://www.github.com/ metrics-lab/ST-fMRI/.

by combining nodes with common functions and large contiguous areas. In this latter configuration, nodes time series correspond to the average BOLD signal in each ROI. Subject-specific ICA nodes and group-averaged 22 ROIs time series were normalised by z-scores.

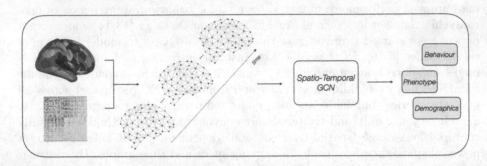

Fig. 1. Illustration of the pipeline used in this study. Group-ICA parcellation regressed into individuals is combined with FC to build a spatio-temporal graph sequence representing time courses of brain activation.

Model. In this paper, long-range, spatio-temporal dynamics of FC are modelled by extending the ST-GCN model [29] (utilised in [5]) to incorporate information over multiple spatial and temporal scales, by adapting the multi-scale, spatio-temporal graph operator (MS-G3D) approach of [16]. Following the ST-GCN model of FC used by [5], functional connectomes are modelled as graphs $\mathcal{G} = (\mathcal{V}, \mathcal{E})$, with $\mathcal{V} = \{v_1, v_2....v_N\}$ representing the set of functional regions (or nodes); and $\mathcal{E}$ representing spatial and temporal connections (or edges). In this framework, the spatial graph $\mathbf{A} \in \mathbb{R}^{N \times N}$:

$$\mathbf{A}_{i,j} = \begin{cases} corr(i,j) \text{ if } i \neq j \\ 0 \text{ otherwise} \end{cases} \tag{1}$$

represents static FC estimated over all time, with $corr(i,j)$ measuring the magnitude of correlation over the full timeseries (1200 frames). Spatio-temporal interactions are then modelled in the form of a feature tensor $\mathbf{X} \in \mathbb{R}^{T \times N \times C}$: $\mathcal{X} = \{x_{n,t} \in \mathbb{R}^C | t, n \in \mathbb{Z}, 1 \leq t \leq T, 1 \leq n \leq N\}$, where the channel dimension of the input feature tensor represents the node time courses ($C = 1$). The MS-G3D framework [16] models spatio-temporal graph convolutions (ST-GC) as blocks composed of two streams: a *factorised pathway* which decouples space and time, and a *G3D pathway* which aggregates space-time information. In *the factorised pathway*, space-only convolutions are represented by:

$$\mathbf{X}_t^{(l+1)} = \sigma\left(\tilde{\mathbf{D}}^{-1/2}(\mathbf{A} + \mathbf{I})\tilde{\mathbf{D}}^{-1/2}\mathbf{X}_t^{(l)}\Theta^{(l)}\right) \tag{2}$$

with $\Theta^{(l)}$ the learnable weights at layer l and $\tilde{\mathbf{D}}$ the degree matrix of $(\mathbf{A} + \mathbf{I})$. Here, σ is a ReLU activation function, except for the last ST-GC block of the model. Still in *the factorised pathway*, temporal-only convolutions (TCN) are implemented with a bottleneck design, including four convolutional filters, all with kernels of size (3, 1) along the temporal dimension but with various dilation factors ($d \in [1,4]$), and residual connections. In parallel, the *G3D pathway* splits the full frame sequence into sub time-windows of $\tau \in \{3,5\}$, and combines regional spacetime information using a bespoke spatio-temporal graph convolution operation (G3D), applied on every single sub time-window:

$$\left[\mathbf{X}_{(\tau)}^{(l+1)}\right]_t = \sigma\left(\tilde{\mathbf{D}}_{(\tau)}^{-1/2}\tilde{\mathbf{A}}_{(\tau)}\tilde{\mathbf{D}}_{(\tau)}^{-1/2}\left[\mathbf{X}_{(\tau)}^{(l)}\right]_t \Theta^{(l)}\right) \tag{3}$$

with

$$\tilde{\mathbf{A}}_{(\tau)} = \begin{bmatrix} (\mathbf{A}+\mathbf{I}) & \dots & (\mathbf{A}+\mathbf{I}) \\ \vdots & \ddots & \vdots \\ (\mathbf{A}+\mathbf{I}) & \cdots & (\mathbf{A}+\mathbf{I}) \end{bmatrix} \in \mathbb{R}^{\tau N \times \tau N} \tag{4}$$

The G3D operation (Eq. (3)), therefore extends the spatial convolution (Eq. (2)) into a space-time operation by stacking the adjacency matrix $(\mathbf{A}+\mathbf{I})$ into $\tilde{\mathbf{A}}_{(\tau)}$ connecting nodes across τ frames, in space and time: edges being extended to connect nodes to their neighbours across the entire sub time-window; in this way constructing a spatio-temporal subgraph $\mathcal{G}_{(t)} = (\mathcal{V}_{(t)}, \mathcal{E}_{(t)})$. $\tilde{\mathbf{D}}_{(\tau)}$ and $\mathbf{X}_{(\tau)}$ are natural extensions of the previously introduced notations.

In addition, a multi-scale (MS) operator can be embedded into any of the previously described convolutions. This aims at aggregating more structural information, by relating a node to its K-closest neighbours, through substituting the convolution operation in (Eq. 2) by $\sum_{k=0}^{K} \tilde{\mathbf{D}}_{(k)}^{-1/2}\tilde{\mathbf{A}}_{(k)}\tilde{\mathbf{D}}_{(k)}^{-1/2}\mathbf{X}_t^{(l)}\Theta_{(k)}^{(l)}$ where:

$$\tilde{\mathbf{A}}_{(k)} = \mathbf{I} + \mathbb{1}\left(\tilde{\mathbf{A}}^k \geq 1\right) - \mathbb{1}\left(\tilde{\mathbf{A}}^{k-1} \geq 1\right) | \tilde{\mathbf{A}}_{(1)} = \mathbf{A} + \mathbf{I}; \tilde{\mathbf{A}}_{(0)} = \mathbf{I} \tag{5}$$

In summary, where the ST-GCN model [29] adopted by [5] factorises spatio-temporal convolutions into spatial-only and temporal-only blocks, useful for modelling information flow along space and time (but limited in their capacity to model space-time interactions), our method adds in parallel a space-time convolution module connecting all nodes in a short temporal window. Importantly, the TCN block adopted here from MS-G3D [16] also improves capacity for more long-range temporal modelling by aggregating multi-scale temporal convolutions, over several temporal scales.

Architecture. Network implementation (illustrated in Fig. 2) follows the design principles of MS-G3D [16] but adapts it to the case of FC. At each training iteration, a time window of T consecutive frames is sampled at random from the entire sequence and is passed as input to the model. In the present architecture (Fig. 2), three ST-GC blocks were stacked with 96, 192, 384 outputs channels, respectively. In each of these blocks, two *G3D pathways* convolutions (Eq. (3)) were implemented; these were passed spatio-temporal graphs formed from time-windows of consecutive frames *at two temporal lengths* ($\tau \in \{3,5\}$). In parallel, the factorised pathway applies spatial-only GCN (Eq. (2)), followed by temporal-only convolutions TCN, on the entire time-window sequence. Then, a multi-scale temporal module aggregates the two pathways, increasing long-range temporal modelling by connecting regional sub-windows. Finally, predictions are made either as classification or regression, by simply replacing a sigmoid activation layer (for classification) with a regression head for the fluid intelligence task. We refer to this architecture as *Brain-MS-G3D*.

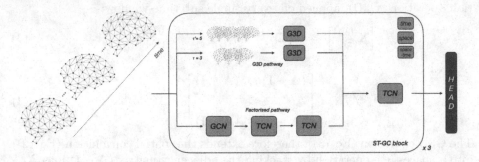

Fig. 2. Architecture of our Brain-MS-G3D. A sequence of T frames is used as input of the model. Each G3D pathway processes T sub time-windows of length $\tau \in \{3, 5\}$, while the factorised pathway process the entire sequence T with spatial-only (GCN) and temporal-only (TCN) blocks. A final TCN block aggregates all pathways.

Experimental Setup. We first trained models, the ST-GCN [5] and our Brain-MS-G3D, on the 22 ROIs group-average cortical parcellation used by [5]. Then both models were also compared on the subject-specific node time series derived from dual-regressed group-ICA at different dimensionalities (15 to 200 nodes). Following the approach used in [5], we benchmark the proposed MS-G3D [16] model against the ST-GCN [29] used in [5], by training models using 5-fold cross-validation. In both cases models were trained on different length time-windows T, from the entire 1200 time-points sequence. At each training iteration, a temporal window was randomly extracted from the entire sequence to be processed by the model. At inference time, predictions from $V = 64$ voters were averaged. In all cases, an Adam optimiser was used with binary cross-entropy loss (for classification) and mean squared error (MSE) for regression; learning rate was set to $1e^{-3}$ by default and with a weight decay of $1e^{-3}$. All experiments were run on a single NVIDIA TITAN RTX 24GB GPU. The batch size was maximised to fit into memory depending on the node resolution: for instance, 32 elements per batch for 200 nodes but 512 for 25 nodes.

4   Results

Results obtained are summarised in Fig. 3 and Table 1. In Table 1, we compared our method to the ST-GCN model [5], on the 22 ROIs and the ICA subject-specific node timeseries at different dimensionalities for sex classification and fluid intelligence prediction, and to results reported in a recent method which models FC dynamics using spatio-temporal attention on graphs [10]. In Fig. 3, performance on sex classification is evaluated across different ICA (and group parcellation) dimensionalities, on models trained and optimised for a maximum number of 2000 iterations. Results show that our Brain-MS-G3D model, inspired by [16], outperforms the ST-GCN model [29], used in [5], at every dimensionality, with an average increase in performance of +6.8%, and a maximum of +11.1% at dimensionality 100. In Table 1, we report results for optimised trainings with more iterations, using the best time-window length obtained from Fig. 3. Our Brain-MS-G3D notably reached a 94.4% sex classification accuracy at dimensionality 200. We also note in Table 1 an increase in classification for both models when moving from the group-average parcellation (22 ROIs) to a similar ICA dimensionality (25 nodes). For sex classification, we experimented the impact of changing the length of time-windows T for three node dimensionalities (15, 25, 50): with $T \in \{50, 75, 100\}$, representing respectively 4.2%, 6.3% and 8.3% of the entire time sequence. Range of performance across time windows is reported through the vertical bars in Fig. 3, where plotted values correspond to the highest accuracy for $T \in \{50, 75, 100\}$. In the case of Brain-MS-G3D, all experiments suggest an increase in performance with larger time-windows (best results with $T = 100$), while the gap reduces when spatial resolution increases. By comparison, the ST-GCN model demonstrates a larger variance in performance, and the 100-time window network always under-performing the others. As widely reported in the literature [11,12,19], fluid intelligence was a much more difficult task to predict. As detailed in [20], performances for fluid intelligence performance are reported as correlation values in Table 1. Our results quantitatively outperform established methodology such as MAGE and HMM on fluid intelligence prediction (see [19,26]).

5 Discussion

While our approach introduced a spatio-temporal modelling of functional connectivity that has proved to increase phenotype predictions, additional experiments revealed that the multi-scale operator (Eq. (5)), used in [16], has only a small impact on Brain-MS-G3D performances. In the case of brain connectomes, the spatial graph is by construction a weighted-undirected graph that connects all nodes together, while in skeleton analysis the spatial graph is much sparser. Thresholding correlation values in the adjacency matrix could increase the impact of spatial scaling (Eq. (5)). Additionally, in Fig. 3, we illustrated that increasing the spatial resolution of the ICA tends to improve performances. However, the length of time sequence has also been pointed as crucial to understand the temporal hierarchy of FC [27]. We had to limit the study to a relatively small number of time frames per sequence (maximum of 128), mostly due to hardware limitations, but experiments on larger time windows with smaller ICA resolution showed dramatic increases in prediction accuracy, which should be further investigated. Finally, Fig. 4 illustrates the impact of parcellation for modelling subject-variability in functional connectivity: atypical subject-specific activation might be averaged out by parcellation.

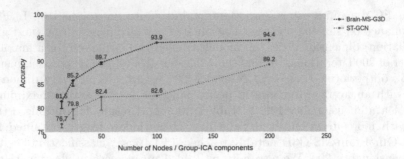

Fig. 3. Comparison of sex classification results between ST-GCN and our Brain-MS-G3D on ICA subject-specific data at different resolution (15, 25, 50, 100, 200).

Table 1. Results of sex classification and fluid intelligence prediction on the HCP dataset. Results presented were optimised for a maximum of 10k iterations. In square brackets, result reported in original publications.

Methods	Data	Sex (% acc)	Fluid intelligence (corr)
ST-GCN [5]	22 Group ROIs	81.8 [83.7]	0.144
STAGIN-SERO [10]	400 ROIs	[88.20]	N/A
STAGIN-GARO [10]	400 ROIs	[87.01]	N/A
Brain-MS-G3D	22 Group ROIs	**84.7**	0.269
Brain-MS-G3D	ICA-15	81.5	0.286
Brain-MS-G3D	ICA-25	86.1	0.313
ST-GCN [5]	ICA-25	82.1	N/A
Brain-MS-G3D	ICA-50	90.9	**0.325**
Brain-MS-G3D	ICA-100	93.9	0.317
Brain-MS-G3D	ICA-200	**94.4**	0.324

(a) 22 ROIs group-average parcellation derived from [6]

(b) Group-average activation map

(c) Subject-specific activation map

Fig. 4. Pink arrows highlight Area 55b, which shows topographic variability between subject. In (a), the functional connectivity of this region is averaged across the larger ROI (light green). (b) group average functional connectivity of the HCP language task contrast "Story vs. Baseline", and (c) shows atypical functional connectivity of the same language task contrast in a single subject. (Color figure online)

In conclusion, this study improved behavioural and demographic predictions on the HCP dataset by introducing a state-of-the-art method for skeleton-based action recognition to explicitly modelling spatio-temporal dynamics of FC, and accounting for inter-subject variability in functional organisation through the use of dual-regressed ICA maps. Future work will explore the potential for improving performance further by increasing the spatial resolution of the model.

Acknowledgments. Data were provided by the Human Connectome Project, WU-Minn Consortium (Principal Investigators: David Van Essen and Kamil Ugurbil; 1U54MH091657) funded by the 16 NIH Institutes and Centers that support the NIH Blueprint for Neuroscience Research; and by the McDonnell Center for Systems Neuroscience at Washington University [25].

References

1. Beckmann, C., Smith, S.: Probabilistic independent component analysis for functional magnetic resonance imaging. IEEE Trans. Med. Imaging **23**(2), 137–152 (2004)
2. Bijsterbosch, J.D., Woolrich, M.W., Glasser, M.F., Robinson, E.C., Beckmann, C.F., et al.: The relationship between spatial configuration and functional connectivity of brain regions. Elife (2018)
3. Dsouza, N.S., Nebel, M.B., Crocetti, D., Robinson, J., Mostofsky, S., et al.: M-GCN: a multimodal graph convolutional network to integrate functional and structural connectomics data to predict multidimensional phenotypic characterizations (2021)
4. Finn, E.S., Shen, X., Scheinost, D., Rosenberg, M.D., Huang, J., et al.: Functional connectome fingerprinting: identifying individuals using patterns of brain connectivity. Nat. Neurosci. **18**(11), 1664–1671 (2015)
5. Gadgil, S., Zhao, Q., Pfefferbaum, A., Sullivan, E.V., Adeli, E., et al.: Spatio-temporal graph convolution for resting-state fMRI analysis (2021)
6. Glasser, M.F., Coalson, T.S., Robinson, E.C., Hacker, C.D., Harwell, J., et al.: A multi-modal parcellation of human cerebral cortex. Nature **7615**, 171–178 (2016)
7. Glasser, M.F., Sotiropoulos, S.N., Wilson, J.A., Coalson, T.S., Fischl, B., et al.: The minimal preprocessing pipelines for the human connectome project. NeuroImage **80**, 105–124 (2013)
8. Harrison, S.J., Bijsterbosch, J.D., Segerdahl, A.R., Fitzgibbon, S.P., Farahibozorg, S.R., et al.: Modelling subject variability in the spatial and temporal characteristics of functional modes. NeuroImage **222**, 117226 (2020)
9. Huang, Z.A., Zhu, Z., Yau, C.H., Tan, K.C.: Identifying autism spectrum disorder from resting-state fMRI using deep belief network. IEEE Trans. Neural Netw. Learn. Syst. (2020)
10. Kim, B.H., Ye, J.C., Kim, J.J.: Learning dynamic graph representation of brain connectome with spatio-temporal attention (2021)
11. Kong, R., Li, J., Orban, C., Sabuncu, M.R., Liu, H., et al.: Spatial topography of individual-specific cortical networks predicts human cognition, personality, and emotion. Cereb. Cortex **29**(6), 2533–2551 (2019)
12. Kong, R., Yang, Q., Gordon, E., Xue, A., Yan, X., et al.: Individual-specific areal-level parcellations improve functional connectivity prediction of behavior. bioRxiv (2021)

13. Ktena, S.I., Parisot, S., Ferrante, E., Rajchl, M., Lee, M., et al.: Metric learning with spectral graph convolutions on brain connectivity networks. NeuroImage **169**, 431–442 (2018)
14. Li, X., Zhou, Y., Gao, S., Dvornek, N., Zhang, M., et al.: BrainGNN: interpretable brain graph neural network for fMRI analysis. bioRxiv (2020)
15. Liégeois, R., Li, J., Kong, R., Orban, C., Van De Ville, D., et al.: Resting brain dynamics at different timescales capture distinct aspects of human behavior. Nat. Commun. **10**(1), 1–9 (2019)
16. Liu, Z., Zhang, H., Chen, Z., Wang, Z., Ouyang, W.: Disentangling and unifying graph convolutions for skeleton-based action recognition (2020)
17. Marquand, A.F., Kia, S.M., Zabihi, M., Wolfers, T., Buitelaar, J.K., et al.: Conceptualizing mental disorders as deviations from normative functioning. Mol. Psychiatry **10**, 1415–1424 (2019)
18. Marquand, A.F., Rezek, I., Buitelaar, J., Beckmann, C.F.: Understanding heterogeneity in clinical cohorts using normative models: beyond case-control studies. Biol. Psychiatry **80**(7), 552–561 (2016)
19. Pervaiz, U., Vidaurre, D., Gohil, C., Smith, S.M., Woolrich, M.W.: Multi-dynamic modelling reveals strongly time-varying resting fMRI correlations. bioRxiv (2021)
20. Pervaiz, U., Vidaurre, D., Woolrich, M.W., Smith, S.M.: Optimising network modelling methods for fMRI. NeuroImage **211**, 116604 (2020)
21. Robinson, E.C., Garcia, K., Glasser, M.F., Chen, Z., Coalson, T.S., et al.: Multimodal surface matching with higher-order smoothness constraints. NeuroImage **167**, 453–465 (2018)
22. Robinson, E.C., Jbabdi, S., Glasser, M.F., Andersson, J., Burgess, G.C., et al.: MSM: a new flexible framework for multimodal surface matching. NeuroImage **100**, 414–426 (2014)
23. Smith, S.M., Nichols, T.E., Vidaurre, D., Winkler, A.M., Behrens, T.E., et al.: A positive-negative mode of population covariation links brain connectivity, demographics and behavior (2015)
24. Smith, S.M., Vidaurre, D., Beckmann, C.F., Glasser, M.F., Jenkinson, M., et al.: Functional connectomics from resting-state fMRI. Trends Cogn. Sci. **17**(12), 666–682 (2013)
25. Van Essen, D.C., Smith, S.M., Barch, D.M., Behrens, T.E., Yacoub, E., et al.: The WU-Minn human connectome project: an overview. NeuroImage **80**, 62–79 (2013)
26. Vidaurre, D., Abeysuriya, R., Becker, R., Quinn, A.J., Alfaro-Almagro, F., et al.: Discovering dynamic brain networks from big data in rest and task. NeuroImage **180**, 646–656 (2018)
27. Vidaurre, D., Smith, S.M., Woolrich, M.W.: Brain network dynamics are hierarchically organized in time. Proc. Natl. Acad. Sci. **114**(48), 12827–12832 (2017)
28. Wolfers, T., Rokicki, J., Alnæs, D., Berthet, P., Agartz, I., et al.: Replicating extensive brain structural heterogeneity in individuals with schizophrenia and bipolar disorder. Hum. Brain Mapp. **42**(8), 2546–2555 (2021)
29. Yan, S., Xiong, Y., Lin, D.: Spatial temporal graph convolutional networks for skeleton-based action recognition (2018)

H3K27M Mutations Prediction for Brainstem Gliomas Based on Diffusion Radiomics Learning

Ne Yang[1], Xiong Xiao[2], Xianyu Wang[1], Guocan Gu[2], Liwei Zhang[2], and Hongen Liao[1(✉)]

[1] Department of Biomedical Engineering, School of Medicine,
Tsinghua University, Beijing, China
liao@tsinghua.edu.cn
[2] Department of Neurosurgery, Beijing Tiantan Hospital, Capital Medical
University, Beijing, China

Abstract. H3K27M mutation is the most common mutation in brainstem gliomas (BSGs), which is related with highly invasive neoplasms and poor prognosis. Accurate presurgical and noninvasive prediction of H3K27M mutations based on preoperative multi-modal neuroimaging is of great clinical value in the diagnosis, prognosis and therapeutic selection of BSGs. Traditional BSG radiomics models usually only focus on tumor local morphometric characteristics. However, given that highly invasive BSGs may significantly affect large-scale brain network connectivity, we reasonably infer that local radiomics and global connectomics may provide different perspectives for H3K27M genotype prediction. Therefore, we define a graph-based diffusion radiomics learning model to integrate these two kinds of features seamlessly. Specifically, edges of the defined brain network are determined by neural fiber connections, while node features of brainstem are governed by local tumor radiomics. Upon this model, we further propose a multi-mechanism diffusion convolutional network to couple multi-modal information and generate a joint representation for brain disease diagnosis. By graph diffusion convolution, the local radiomics information spread along the brain network structure to enhance graph representation learning, and eventually the learned diffusion radiomics features contribute to disease prediction. Experiments on a real BSG dataset demonstrate the effectiveness and advantages of our proposed method for preoperative prediction of H3K27M statuses.

Keywords: Brainstem glioma · Brain networks · Radiomics · Deep learning · Graph neural network

1 Introduction

Brainstem gliomas (BSGs) are a group of highly heterogeneous tumors that occur in the brainstem [1]. Recently, some findings indicate that the H3K27M mutation, which is

---

Electronic supplementary material The online version of this chapter (https://doi.org/10.1007/978-3-030-87586-2_16) contains supplementary material, which is available to authorized users.

related with highly invasive neoplasms, can serve as a qualified biomarker for differential diagnosis, prognosis assessment and treatment strategy selection for BSG patients [2, 3]. However, the H3K27M mutation status information is obtained through surgical biopsy traditionally, which limited the extensive clinical implementation of H3K27M. Therefore, non-invasive and preoperative prediction of H3K27M statuses for BSGs is highly desired, which is convenient, cost-saving and of great significance for a better treatment plan.

Radiomics is an emerging method that examines the relationship between the medical imaging and molecular indicators based on artificial intelligence [4]. Recently, a few radiomics studies have been carried out to predict BSG relative molecular biomarkers statuses based on the tumor characteristics extracted from preoperative structural MRI [5, 6, 12]. A major limitation of the traditional radiomics is only considering the local morphological features of the tumor area. Some findings indicate that BSGs, which have diffusive infiltration along the white matter, may severely affect large-scale brain network connectivity [7]. Recently, brain network provides a powerful representation of human brain, offering new insights for neuroscientists to identify brain diseases [8]. Some previous works have found that the connectomics extracted from the whole brain network can sensitively reflect the differences between different molecular types [9, 10]. Considering that both the tumor appearance and brain connections could be associated with genotypes and molecular indicators, we reasonably infer that local radiomics and global connectomics features may provide different perspectives for BSG preoperative diagnosis. Deep learning methods have been successfully applied to neuroimaging data analysis [11, 12]. However, how to effectively fuse these two kinds of multi-modal information for brain disease analysis still remains a challenge.

To tackle this problem, we define a graph-based diffusion radiomics model to integrate these two kinds of features seamlessly. The full pipeline of the proposed framework is shown in Fig. 1. Generally, a brain network can be characterized as a graph by a series of nodes and edges, and the graph-level task can be solved by graph neural networks (GNNs) [13]. GNN is the state-of-the-art research field of deep graph structure learning, which can improve the performance of various network neuroscience tasks [14–16]. In our model, the edges are determined by fiber tractography extracted from diffusion MRI (dMRI) and the nodes are defined by the physiological template, while node features of brainstem are governed by tumor radiomics features extracted from structural MRI (sMRI). Upon the designed graph, we further propose a deep graph learning framework called Multi-Mechanism Diffusion-Convolutional Neural Network (MMDC-NN) to couple multi-modal information for BSG preoperative diagnosis through graph diffusion convolution [17], instead of using each individual modality alone. By graph diffusion, the model achieves a deep fusion of local radiomics and global connectomics to learn the diffusion radiomics.

Compared with the existing works for BSG preoperative diagnosis, ours proposed a novel radiomics analysis model, which is the first to give the prediction of H3K27M mutation from the perspective of brain networks using deep graph learning. Our main contributions include (1) seamlessly integrating multi-modal information via a graph-based diffusion radiomics model, (2) proposing a multi-mechanism diffusion convolution kernel to learn more informative latent representation from diffusion process, and

(3) discovering potential BSG-related biomarkers based on a node-level self-attention pooling layer, through the supervised graph-level classification. Our proposed method is verified on a real BSG dataset and demonstrates promising performance via an extensive comparison with the state-of-the-art methods for BSG preoperative diagnosis.

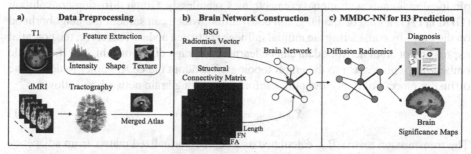

Fig. 1. Pipeline of Diffusion-Radiomics. First, the sMRI-based radiomics features are extracted from tumors, and the dMRI-based tractography are parcellated by a merged atlas (a). Then, the multi-modal information is coupled to construct brain networks (b). Finally, the graphs are fed into our proposed MMDC-NN to encode diffusion radiomics, which performs H3K27M mutation prediction and derives a brain saliency map associated with gene expression (c).

2 Proposed Method

2.1 Materials and Preprocessing

This study was approved by the research-ethics committee of Beijing Tiantan Hospital. Totally, 109 BSG patients with available preoperative 3D-T1 and dMRI images, and available information about H3K27M mutation status were enrolled, consisting of 62 patients of H3K27M mutation status and 47 patients with wild-type H3K27M.

3DSlicer 4.8 [18] was used to segment the BSG in each T1 image by two neuro-surgeons manually. Then a total number of 107 local radiomics features were extracted from the tumor region using PyRadiomics 2.2.0 library [19], including shape, intensity and texture features. The dMRI data were processed by using the PANDA 1.3.1 suite [20]. After deterministic tractography, the SC networks were constructed based on a merged atlas of Automated Anatomical Labeling (AAL) atlas [21] and Harvard-Oxford Atlas (HOA) [22], including 132 regions of interest (ROIs). Then we calculated three types of SC as average FA along the fibers, total number of fibers, and average length of fibers between two ROIs. More details will be available in supplementary materials.

2.2 Problem Formulation

A brain structural network can describe the interaction between paired brain regions using a graph structure $G = \{V, E\}$. The node set $V = \{v_i\}_{i=1}^{N}$ indicating brain regions defined by the merged atlas, which is collectively described by an $N \times F$ matrix $X =$

$[x_1, \ldots, x_N]^T$ of features. The edge set $E = \{e_{ij}\}$ are encoded by an $N \times N$ adjacency matrix A, where a_{ij} is the corresponding weight of SC between ROI i and j. Specifically, the node features of brainstem are governed by BSG radiomics features.

Given that highly invasive BSGs may significantly affect brain connectomics along the white matter tracts from brainstem, we propose a diffusion radiomics model to couple local radiomics with global connectomics seamlessly. Graph diffusion convolution (GDC) [17] is a kind of spatial graph convolution, which can smooth the neighborhood on the graph by considering the mutual influence between nodes. Therefore, the potential expression of brain network data can be learned via graph diffusion which perform node embedding with respect to the graph topology. We adopt the GDC kernel to design the diffusion process of tumor radiomics information along brain network as follows:

$$X_{\text{diffusion}} = f(W \odot PX) \tag{1}$$

The transmission matrix P is calculated as a degree-normalized matrix from adjacency matrix A, and p_{ij} represents the transmission probability from ROI i to ROI j in a one-step graph diffusion. In this way, a brain structural network can be interpreted as a random-walk net where tumor-relevant biological information flows from node to node. We then design a novel multi-mechanism GDC kernel, to obtain the potential whole brain feature representation $X_{\text{diffusion}} \in \mathcal{R}^{N \times F}$ for BSG preoperative diagnosis.

2.3 Multi-Mechanism Diffusion-Convolution (MMDC)

Given a target node v_i, the traditional GDC kernel first updates the node features x_i by aggregating feature information from its immediate neighbors $\mathcal{N}_i$ (including v_i itself):

$$x_i^{diffusion} = \text{ReLU}\left(W \odot \sum_{v_j \in \mathcal{N}_i} p_{ij} \cdot x_j\right) \tag{2}$$

Here, ReLU is a non-linear activation, $W \in \mathcal{R}^{F' \times F}$ is the learnable model parameter of the fully-connected (FC) layer, and $\odot$ represents the element-wise multiplication.

It is worth noting that the transmission weight p_{ij} in the above formula is fixed, which may lead to poor generalization of node's hidden representation. Though these predefined values from SC matrix could reflect the brain network physical connections, they might not be optimal for the diffusion-radiomics learning in our research. Therefore, the dynamic adjustment of the transmission matrix P during graph diffusion learning is necessary. We adopt the idea of graph attention networks (GATs) [23] to learn dynamic edge weights:

$$p_{ATT,ij} = \frac{\exp\left(\text{LeakyReLU}\left(\vec{a}^T[Wx_i||Wx_j]\right)\right)}{\sum_{v_k \in \mathcal{N}_i} \exp\left(\text{LeakyReLU}\left(\vec{a}^T[Wx_i||Wx_k]\right)\right)}, \tag{3}$$

where $\vec{a} \in \mathcal{R}^{2F'}$ and $W \in \mathcal{R}^{F' \times F}$ are model parameters, and $||$ represents concatenation. To the end, we integrate multiple metrics of SC to define a MMDC kernel, which can increase the learning ability of the model. Finally, the transmission matrix P_{MMDC} and the whole diffusion convolution on the feature matrix X can be like:

$$P_{MMDC} = \alpha P_{FA} + \beta P_{FN} + (1 - \alpha - \beta) P_{\text{Length}} + P_{ATT}, \tag{4}$$

$$x_i^{diffusion} = \text{ReLU}\left(W \odot \sum_{v_j \in \mathcal{N}_i} x_j \cdot p_{ij,MMDC} \right), \tag{5}$$

where α and β are learnable parameters balancing different diffusion mechanisms. P_{FA}, P_{FN} and P_{Length} are the pre-defined three different kinds of SC measurement without graph attention. P_{ATT} is the attention transmission matrix for dynamic adjustment.

2.4 Diffusion-Radiomics

To obtain a compact and discriminative representation related to graph classification task, GNNs are often combined with pooling and readout operations [13], to perform feature dimensionality reduction and reduce the size of parameters, which generate smaller graph representations and thus avoid overfitting. The graph pooling layer coarsens each graph into a subgraph for down-sampling, and the readout layer translates node representations of each graph into a graph representation. Finally, by applying a multi-layer perception (MLP) to graph representations, we can perform graph classification. The architecture of proposed MMDC-NN is illustrated in Fig. 2.

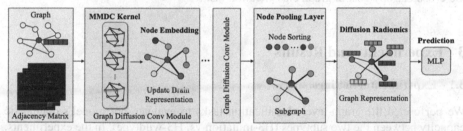

Fig. 2. MMDC-NN for brain graph classification. Graph-structured brain networks based on tumor radiomics and SC matrix are fed into MMDC-NN for learning diffusion-radiomics.

Node Embedding. Based on our proposed MMDC kernel, we can get the potential brain representation via node embedding to achieve a deep integration of BSG radiomics and brain networks. Previous research proves that a k-hop convolution kernel can be divided into k 1-hop convolutions [24]. Therefore, we stack multiple MMDC layers to enhance node embdeding and improve the ability of the model to capture discriminative features, then the final hidden representation of each node can receive information from diverse ranges of the neighbors.

Node Pooling Layer. Moreover, the last MMDC kernel is followed by a global pooling layer at node-level to generate smaller representations and thus avoid overfitting and perform node selection. Most recently, the SAGPool approach is proposed [25], which learns the pooling in a self-attention manner based on both the node features and graph topology. The choices of keeping which nodes are determined based on the node importance score S, which is obtained by graph convolution as follows:

$$S = \sigma(GCN(X, A)) \tag{6}$$

Here, A is binarized and used to determine if there is a connection between two nodes. Moreover, to consider the two-hop connection, an edge augmentation of SAGPool is proposed to involve the information of two-hop neighbors:

$$S = \sigma\left(GCN\left(X, \left(A + A^2\right)\right)\right) \tag{7}$$

Then, the node-level pooling retains some of the nodes in the input graph:

$$\text{idx} = \text{toprank}(S, \lceil kN \rceil), \ X_{out} = X_{idx,:} \odot S_{idx}, \ A_{out} = A_{idx,idx} \tag{8}$$

Here, k is a pooling ratio that determines the number of nodes to keep, and the "toprank" function finds the indices corresponding to the largest $\lceil kN \rceil$ elements in score vector S.

Decision Module. Finally, a readout layer summarizes the final graph-level representation based on node hidden representations. We use $Z = \oplus\{X_{out,i=1,...,N_{selected}}\}$ to generate a representation of the entire graph, where $\oplus$ operates concatenation element-wisely. Then Z is sent to a MLP to give the final prediction, and the model is trained to minimize the cross-entropy cost of graph classification task.

3   Experiments and Results

3.1   Experimental Settings

We performed the graph-level classification tasks to evaluate our model's diagnostic capacity between the two subtypes (H3-mutation vs. H3-wildtype). In the experiments, we set the parameters N = 132, F = 107. The model architecture was implemented with 2 MMDC layers with parameter $F^{'(1)} = 32$, $F^{'(2)} = 16$, followed by a pooling layer kept the top 10% important nodes. Then a 3-layer MLP took the final graph-level representation $Z \in \mathcal{R}^{16}$ as input and predicted the patient's subtype.

To validate the effectiveness of our proposed method, we also compared it with 5 competing methods for H3K27M genotype prediction. According to the input of the model, these methods can be divided into three types: Radiomics based methods (i.e., Radiomics-SVM [5], and CNN-SVM [12]), Connectomics based methods (i.e., Connectomics-SVM [9], and BrainNetCNN [11]), and Multi-modal based methods (i.e., Concatenation-SVM [9], and MMDC-NN (ours)). CNN-SVM, BrainNetCNN and MMDC-NN use deep models, while the rest are traditional machine learning methods. The performance was evaluated via 5-fold cross validation. The average accuracy (ACC), average sensitivity (SEN), average specificity (SPE) and the average area under receiver operating characteristic curve (AUC) were reported as the final performance measures.

3.2 H3K27M Mutation Prediction Results

As shown in Table 1, our diffusion-radiomics model achieves the best performance for BSG preoperative diagnosis, i.e., H3K27M mutation prediction task. First, we evaluated H3K27M status prediction performance using features from single modality, i.e., based on radiomics and connectomics features, separately. We can see that, radiomics-based methods achieve higher accuracy than connectomics-based methods, while connectomics show higher specificity (i.e., the proportion of H3-wildtype that are correctly predicted). Second, considering that different modalities could provide complementary information and thus may enhance the prediction performance, we performed experiments based on multi-modal fusion in two ways, (1) construct a new feature matrix by simply concatenating radiomics and connectomics features, (2) learn diffusion-radiomics through our proposed MMDC-NN. As can be seen, the modality fusion can help improve the prediction performance in terms of accuracy and sensitivity (i.e., the proportion of H3-mutation that are correctly predicted), indicating the benefits of integrating local radiomics and large-scale connectomics. Specifically, it is worth noting that the accuracy of our MMDC-NN is higher than concatenation-based methods, which means that the diffusion radiomics generated by our modal could improve the discriminative power of the original feature by achieve a deep fusion of local radiomics and global connectomics information.

Table 1. Comparison with competing methods.

Method	ACC (%)	AUC	SEN (%)	SPE (%)
Radiomics-SVM [5]	82.8 ± 2.1	0.843	80.3 ± 2.2	82.2 ± 1.4
CNN-SVM [12]	83.2 ± 1.6	0.840	83.4 ± 2.0	84.1 ± 2.0
Connectomics-SVM [9]	81.4 ± 1.4	0.859	79.6 ± 2.6	**85.0 ± 1.6**
BrainNetCnn [11]	79.3 ± 1.8	0.821	81.2 ± 2.1	84.4 ± 2.1
Concatenation-SVM [9]	83.6 ± 1.6	0.863	85.2 ± 2.8	84.2 ± 2.3
MMDC-NN (ours)	**87.9 ± 1.3**	**0.905**	**90.3 ± 2.4**	84.8 ± 1.9

Table 2. Model variations discussion.

Model variations	ACC (%)	AUC	SEN (%)	SPE (%)
FA-DC-NN	85.4 ± 1.3	0.864	88.0 ± 2.0	84.0 ± 1.4
FN-DC-NN	84.8 ± 1.4	0.880	86.6 ± 1.9	84.5 ± 1.3
Length-DC-NN	83.0 ± 1.2	0.856	85.3 ± 2.0	82.4 ± 1.4
Without GAT	84.3 ± 1.5	0.849	87.2 ± 2.1	83.5 ± 1.6
Without SAGPool	82.6 ± 1.3	0.824	84.2 ± 2.7	82.7 ± 2.1
MMDC-NN	**87.9 ± 1.3**	**0.905**	**90.3 ± 2.4**	**84.8 ± 1.9**

Moreover, to explore influence of each component in our model, 5 variant models of MMDC-NN were evaluated as an ablation study. As shown in Table 2, removing the GAT mechanism could cause a significant decrease in performance, while adding the SAGPool could further improve the performance of classification by feeding the decision module with powerful and low-dimension features. We also separately evaluated the learning ability of each SC measurement on diffusion, while the GAT and SAGPool settings were the same as MMDC-NN. As can be seen, integrating multiple SC metrics in graph diffusion can help improve the prediction performance.

3.3 Node Pooling Interpretation

In addition, the most important brain regions affecting the H3K27M prediction are shown in Fig. 3. By averaging the node importance scores among all the subjects, the group-wise significant ROIs associated with BSG diffusion radiomics spread at cerebellum and cerebral. Specifically, the involved cortical areas include the frontal and occipital pole, middle temporal gyrus, precentral and postcentral gyrus. All those regions play vital roles in motor, sense and emotion controls. It is worth noting that our MMDC-NN also highlights some subcortical areas such as amygdala and putamen. The amygdala plays a central role in emotion recognition [26], and the findings of this study may help to further explain the mechanism of emotional personality changes in BSG patients.

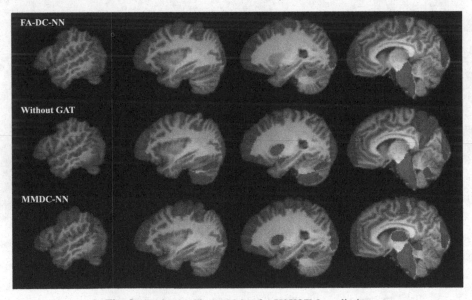

Fig. 3. Brain significance Map for H3K27M prediction

4   Conclusion

In this paper, a novel BSG preoperative diagnosis framework based on a deep graph learning is proposed, which achieves a deep fusion of local radiomics and global connectomics in gene prediction work via diffusion radiomics. In our proposed MMDC-NN, the node embedding is generated by a multi-mechanism graph diffusion, and is supervised by the prediction tasks to detect significant brain regions for disease diagnosis. Our model not only performs better than alternative methods on H3K27M mutation prediction, but also gives interpretable results revealing the significant ROIs related to gene expression. Our method has a potential for understanding brain disease based on multi-modal neuroimage. In the future, we will validate our methods on larger benchmark datasets, and we also plan to extend our model to a spatial-temporal model, and apply to other learning tasks such as characterizing the neurodegeneration in Alzheimer's disease.

Acknowledgments. The authors acknowledge supports from Beijing Municipal Natural Science Foundation (7212202), and National Natural Science Foundation of China (82027807, 81771940).

References

1. Laigle-Donadey, F., Doz, F., Delattre, J.-Y.: Brainstem gliomas in children and adults. Curr. Opin. Oncol. **20**, 662–667 (2008)
2. Khuong-Quang, D.-A., et al.: K27M mutation in histone H3.3 defines clinically and biologically distinct subgroups of pediatric diffuse intrinsic pontine gliomas. Acta Neuropathol. **124**, 439–447 (2012)
3. Hashizume, R., et al.: Pharmacologic inhibition of histone demethylation as a therapy for pediatric brainstem glioma. Nat. Med. **20**, 1394–1396 (2014)
4. Aerts, H., et al.: Decoding tumour phenotype by noninvasive imaging using a quantitative radiomics approach. Nat. Commun. **5**, 4006 (2014)
5. Pan, C.-C., et al.: A machine learning-based prediction model of H3K27M mutations in brainstem gliomas using conventional MRI and clinical features. Radiother. Oncol. J. Eur. Soc. Ther. Radiol. Oncol. **130**, 172–179 (2019)
6. Su, X., et al.: Automated machine learning based on radiomics features predicts H3K27M mutation in midline gliomas of the brain. Neuro Oncol. **22**, 393–401 (2020)
7. Hart, M.G., Price, S.J., Suckling, J.: Connectome analysis for pre-operative brain mapping in neurosurgery. Brit. J. Neurosurg. **30**(5), 506–517 (2016)
8. Rubinov, M., Sporns, O.: Complex network measures of brain connectivity: uses and interpretations. Neuroimage **52**, 1059–1069 (2010)
9. Chen, L., et al.: Multi-label nonlinear matrix completion with transductive multi-task feature selection for joint MGMT and IDH1 status prediction of patient with high-grade gliomas. IEEE Trans. Med. Imaging **37**(8), 1775–1787 (2018)
10. Li, Y., et al.: Less efficient information transfer in Cys-Allele carriers of DISC1: a brain network study based on diffusion MRI. Cereb. Cortex **23**, 1715–1723 (2013)
11. Kawahara, J., et al.: BrainNetCNN: convolutional neural networks for brain networks; towards predicting neurodevelopment. Neuroimage **146**, 1038–1049 (2017)
12. Liu, J., et al.: A cascaded deep convolutional neural network for joint segmentation and genotype prediction of brainstem gliomas. IEEE Trans. Biomed. Eng. **65**(9), 1943–1952 (2018)
13. Wu, Z., Pan, S., Chen, F., Long, G., Zhang, C., Yu, P.: A comprehensive survey on graph neural networks. IEEE Trans. Neural Netw. Learn. Syst. **32**(1), 4–24 (2021)
14. Bessadok, A., Mahjoub, M. A., Rekik, I.: Graph neural networks in network neuroscience. arXiv preprint arXiv:2106.03535 (2021)

15. Zhang, W., Zhan, L., Thompson, P., Wang, Y.: Deep representation learning for multimodal brain networks. In: Martel, A.L., et al. (eds.) MICCAI 2020. LNCS, vol. 12267, pp. 613–624. Springer, Cham (2020). https://doi.org/10.1007/978-3-030-59728-3_60

16. Huang, J., Zhou, L., Wang, L., Zhang, D.: Integrating functional and structural connectivities via diffusion-convolution-bilinear neural network. In: Shen, D., et al. (eds.) MICCAI 2019. LNCS, vol. 11766, pp. 691–699. Springer, Cham (2019). https://doi.org/10.1007/978-3-030-32248-9_77

17. Atwood, J., Towsley, D.: Diffusion-convolutional neural networks. In: Lee, D., Sugiyama, M., Luxburg, U., Guyon, I., Garnett, R. (eds.) Advances in Neural Information Processing Systems. Curran Associates, Inc. (2016)

18. Fedorov, A., et al.: 3D slicer as an image computing platform for the quantitative imaging network. Magn. Reson. Imaging **30**, 1323–1341 (2012)

19. van Griethuysen, J.J.M., et al.: Computational radiomics system to decode the radiographic phenotype. Cancer Res. **77**, e104–e107 (2017)

20. Cui, Z., Zhong, S., Xu, P., He, Y., Gong, G.: PANDA: a pipeline toolbox for analyzing brain diffusion images. Front. Hum. Neurosci. **7**, 42 (2013)

21. Tzourio-Mazoyer, N., et al.: Automated anatomical labeling of activations in SPM using a macroscopic anatomical parcellation of the MNI MRI single-subject brain: Neuroimage **15**, 273–289 (2002)

22. Goldstein, J.M., et al.: Hypothalamic abnormalities in schizophrenia: sex effects and genetic vulnerability. Biol. Psychiatry **61**, 935–945 (2007)

23. Velickovic, P., Cucurull, G., Casanova, A., Romero, A., Liò, P., Bengio, Y.: Graph attention networks. In: International Conference on Learning Representations (ICLR) (2018)

24. Kipf, T., Welling, M.: Semi-supervised classification with graph convolutional networks. In: International Conference on Learning Representations (ICLR) (2017)

25. Lee, J., Lee, I., Kang, J.: Self-attention graph pooling. arXiv preprint arXiv:1904.08082 (2019)

26. Habel, U., et al.: Amygdala activation and facial expressions: explicit emotion discrimination versus implicit emotion processing. Neuropsychologia **45**(10), 2369–2377 (2007)

Constrained Learning of Task-Related and Spatially-Coherent Dictionaries from Task fMRI Data

Sreekrishna Ramakrishnapillai[1]([⊠]), Harris R. Lieberman[3], Jennifer C. Rood[1], Stefan M. Pasiakos[3], Kori Murray[1], Preetham Shankapal[2], and Owen T. Carmichael[1]

[1] Pennington Biomedical Research Center, Baton Rouge, LA, USA
sreekrishna.ramakrishnapillai@pbrc.edu
[2] GE Healthcare – MR Engineering, Bengaluru, India
[3] Military Nutrition Division, US Army Research Institute of Environmental Medicine (USARIEM), Natick, MA, USA

Abstract. Dictionary learning and sparse coding techniques overcome limitations of traditional voxel-level analyses of task-based functional magnetic resonance imaging (fMRI) data by identifying broader temporal and spatial patterns of brain activity. However, prior applications of these methods to task-related fMRI data are not simultaneously optimized to find temporal patterns of activity that change in concert with changes in task conditions and spatial patterns that leverage existing neuroscience knowledge. In this study we present a new sparse dictionary learning method that uses prior knowledge of the temporal pattern of task conditions and the locations of brain regions hypothesized to be involved in the task to decompose fMRI data into temporal patterns of signals that loosely differ between task conditions and sparse spatial patterns that are at least partially similar to known functional network hubs. An efficient on-line optimization framework identifies the temporal and spatial patterns. The method identifies spatial and temporal patterns programmed into synthetic task fMRI data. The proposed method also identifies spatial locations known *a priori* to be activated by the Attention Network Task (ANT) more completely than competing methods when applied to real fMRI data from 20 healthy young individuals aged 18 to 39 years. Simultaneously leveraging the known temporal structure of the task and biasing solutions towards hypothesized network hubs increases the usefulness of sparse dictionary learning methods applied to task fMRI data.

Keywords: Sparse dictionary learning · fMRI · Constrained learning

1 Introduction

A current focus of human neuroscience research is the identification and study of functional brain networks (FBN) that work in a coordinated fashion to execute cognitive tasks [1]. FBNs consist of distributed brain regions ("hubs") each of which has a characteristic temporal pattern of activity. Due to the central role of FBNs in brain development,

A. Abdulkadir et al. (Eds.): MLCN 2021, LNCS 13001, pp. 165–173, 2021.
https://doi.org/10.1007/978-3-030-87586-2_17

aging, and diseases [2], many techniques have emerged that use sparse dictionary learning methods to automatically identify the locations of hubs and temporal patterns, based on fMRI data [3, 4]. This application of sparse dictionary learning methods to FBN identification has been supported by biological findings suggesting that sparse coding is implemented in the brain [5].

To date, most sparse dictionary learning methods for FBN estimation have been applied to resting-state fMRI data and, therefore, do not attempt to identify hubs that are task-engaged (*i.e.*, hubs with temporal patterns of activity that are correlated with changes in task conditions). While resting-state analysis is important, identifying task-engaged FBNs remains one of the central goals of cognitive neuroscience. Supervised dictionary learning is an exception to this rule [6, 7], but this method strictly constrains its temporal patterns to exactly match the temporal patterns of task conditions. It is therefore susceptible to errors when there is even a minor mismatch between the timing of real brain activity relative to task condition timing. In addition, supervised dictionary learning does not leverage prior knowledge about the hypothesized spatial locations of task-related activity. Assisted dictionary learning [8] avoids the problem of strictly adhering to temporal constraints and does not leverage prior knowledge of spatial locations of activity.

This paper presents a method for sparse dictionary learning of FBN spatial and temporal patterns from task fMRI data (Fig. 1) that uses known task condition time courses and hypothesized FBN hub locations to identify FBN hubs that are spatially coherent and task-engaged. We apply the method to synthetic task fMRI data where FBN spatial and temporal properties are known *a priori* and to real task fMRI data where expected FBN hub locations are known from existing research. We compare the performance of the method with that of existing dictionary learning methods designed for fMRI data.

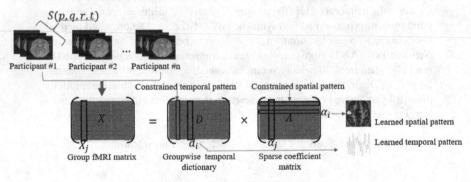

Fig. 1. Dictionary learning of task fMRI data

This paper presents a baseline sparse dictionary learning method as a starting point (Sect. 2). Enhancements that allow the method to leverage known task-related time courses are discussed in Sect. 2.1. We discuss how to make the method leverage hypothesized spatial hub locations and spatial coherence in Sect. 2.2. Optimization within this new framework is described in Sect. 2.3. Section 3 covers experiments on synthetic and real fMRI data.

2   Constrained Online Dictionary Learning

2.1   Dictionary Learning of fMRI Data

Let $S[p, q, r, t]$ denote the fMRI scan of a single participant. Here, p, q, and r are spatial coordinates, and t denotes time. The S from individual j is vectorized into a column vector $X_j \in \mathbb{R}^T$, and all n of these are concatenated to form the group data matrix X (Fig. 1). The aim of sparse dictionary learning is to represent every X as a linear combination of a small number of temporal patterns $\{d_i\}_{i=1}^K$ in a dictionary $D \in \mathbb{R}^{T \times K}$, i.e., $X_j = D\alpha_j$ where $\alpha_j \in \mathbb{R}^K$ represents a spatial pattern within which the temporal patterns are expressed. A is the matrix formed by stacking the column vectors α_j. Typically, $T < K$ and A encouraged to be sparse, i.e., contain a large number of zeroes. We use training signals $\{X_j\}_{j=1}^n$ to learn D and A [9]:

$$\arg\min_{D,A} \sum_{j=1}^n (\|X_j - D\alpha_j\|_2^2 + \lambda \|\alpha_j\|_1^1), \|d_j\|_2 < 1 \; \forall j \tag{1}$$

The first term in the sum encourages low reconstruction error while the second encourages sparsity of α_j. λ controls the tradeoff between these competing terms.

2.2   Incorporating Task Characteristics

Task fMRI is characterized by the imposition of differing experimental conditions on the participant at differing moments in time. Let $(\delta(t))$ represent a time course of task conditions numerically, i.e., $\delta(t) = 0$ (vs. 1) at time t when the first (vs. second) experimental condition is imposed[1]. Given several δ, $\{\delta_1, \delta_2, \ldots \delta_m\}$, we encourage the discovery of temporal patterns that change in concert with changes in experimental conditions by constraining D to the set of matrices C, each with members that contain columns similar to the δ:

$$C \triangleq \left\{ D \in \mathbb{R}^{T \times K} \; \middle| \; \begin{array}{l} \|d_j - \delta_j\|^2 \leq c_\delta, \; j = 1, \ldots, m \\ d_j^T d_j \leq 1, \; j = m+1, \ldots, K \end{array} \right\} \tag{2}$$

Constraining D to the set C requires D to include elements that at least loosely follow the temporal pattern of experimental conditions. Adding this constraint requires us to change the dictionary update step of our optimization algorithm [9] to a two-step process of modifying the constrained and unconstrained temporal patterns separately; otherwise, optimization proceeds as it did without the constraints.

---

[1] Note that the case of two experimental conditions is shown for ease of exposition, but this formulation trivially extends to larger numbers of experimental conditions via multiple δ.

168   S. Ramakrishnapillai et al.

2.3 Constraining Spatial Patterns

Although λ empirically determines the minimum number of voxels that the spatial patterns of FBNs occupy [10], it does not actively encourage biological plausibility of the spatial patterns. To encourage such plausibility, we constrain a subset of the spatial patterns to be similar to those identified as task-engaged in previous fMRI studies. Specifically, for each vectorized spatial pattern α_i, which is a row in A, we generate P_i, a binary version of α_i that has a value of 1 wherever α_i is non-zero. We encourage any *a priori* spatial pattern M to have high spatial overlap with P_i by requiring the Dice coefficient R between P_i and M to be high:

$$R(P_i, M) = \frac{|P_i \cap M|}{|M|} \tag{3}$$

The orthogonal matching pursuit algorithm in the sparse coding step of optimization [9] described in the next section is constrained to incorporate a new Dice-related constraint $\mathfrak{D}_c$ on m designated rows of A. The columns of A are also constrained, as in prior work [10], to be sparse, *i.e.*, to include no more than L nonzero pixels.

$$A \in \mathbb{R}^{K \times n} \ s.t \ \|\alpha_j\|_0^0 \leq L, R(P_i, M) \geq \mathfrak{D}_C, \quad i = 1, \ldots, m \tag{4}$$

2.4 Optimization

Because Eq. (1) is non-convex with respect to D and A jointly, an optimization approach that alternates between D and A is typically utilized, following the Majorization-Minimization (MM) principle [11, 12]. An alternating optimization calculates the sparse code a_j for X_j and updates the temporal patterns in D for a fixed number of iterations over the entire training data [13]. Our work accelerates this traditional approach via an online algorithm inspired by prior work [9]. The representation error $X_j - D\alpha_j$ in (1) is minimized by considering one X_j at a time and updating the temporal patterns based on stochastic block-coordinate descent following prior suggestions that stochastic gradient algorithms provide the fastest solutions among competing approaches for these types of problems [14] while requiring lower speed and less memory than traditional batch optimizers. In addition to its speed advantages, this is a second order optimization method that does not require explicit tuning of a learning rate. This optimizer was implemented on top of the base online optimization code within the SPAMS toolbox [15].

3 Application and Results

3.1 Synthetic fMRI Data Generation Using SimTB

We used SimTB [16] to generate synthetic task-related fMRI data sets. We generated 15 spatiotemporal patterns, each consisting of a single spatial pattern defined on a 100×100 voxel image slice and a temporal pattern of 300 time points (TR = 2s). One of the temporal patterns corresponded to a time course of changing task conditions convolved with a canonical hemodynamic response function (δ) as depicted in Fig. 2, upper right; the other 14 temporal patterns and their corresponding spatial patterns, representing various physiological sources of fMRI signal variability, were taken from a previous publication [8]. We generated 20 synthetic datasets by linearly combining the 15 patterns using randomly generated amplitude scalings of the temporal patterns (with scalings drawn uniformly from [0.5, 1]). Translational head motion in the x and y directions was then drawn uniformly from the range of 0.02 to 0.5 times one voxel size and added to each synthetic scan. Finally, Gaussian amplitude noise at multiple levels (SNR of 10, 20 and 50 dB) was added to create the final synthetic dataset.

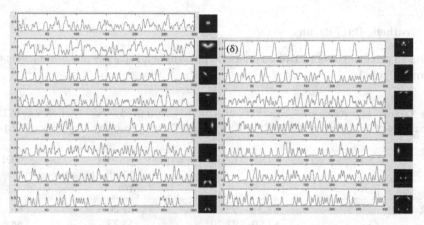

Fig. 2. Spatial patterns and corresponding temporal patterns programmed into the synthetic dataset

3.2 Evaluation of Sparse Dictionary Learning Algorithms

The ground-truth spatial patterns were vectorized into the rows of matrix A_g, and the ground-truth temporal patterns were vectorized into the columns of matrix D_g. We evaluated the proposed algorithm and three competing algorithms in terms of similarity between estimated A and D matrices and these ground-truth matrices. The time course δ was provided as a constraint to the optimizer. The spatial pattern M provided as a constraint to the optimizer is a binarized version of the spatial pattern for one source. For each run of each algorithm, D and A were initialized by randomly drawing entries from a normal distribution (zero mean and variance one) and normalizing columns to unit length. We completed differing runs with differing levels of sparsity L (see Eq. 4), which control the maximum number of nonzero elements in columns of A. Sparsity levels 2, 3, and 4 correspond to 20%, 30% and 40% of the column being nonzero. Within each run, each paired step of optimization consisting of sparse coding and dictionary learning was executed 180 times. We completed 30 runs at each sparsity level for each tested algorithm, with each run applied to a different set of generated synthetic data. Each of the 15 learned temporal patterns in D were correlated with ground-truth time courses in D_g via Pearson correlation. A ground-truth temporal pattern was considered "recovered" if there existed an estimated temporal pattern in D whose Pearson correlation to it was greater than 0.9. Every recovered temporal pattern had a corresponding spatial pattern that had at least 50 percent overlap with the ground-truth spatial pattern. The average number of recovered temporal patterns across all runs at a given sparsity level were then tabulated. We evaluated learning rates γ between 0.001 and 0.25 for K-singular value decomposition (K-SVD) [11] and K-SVD-sparse (K-SVDs); the ones that gave the best results ($\gamma = 0.18$) in terms of maximal correlation between estimated temporal pattern and task condition temporal pattern (δ) was selected. The temporal pattern level sparsity parameter for K-SVDs was set to 10.

3.3 Synthetic Data Results

The mean percentages of recovered temporal patterns at all sparsity and noise levels are shown in Table 1.

Table 1. Average percentage of recovered atoms for the Constrained Dictionary Learning (CDL) compared to other algorithms KSVD [11], KSVDs, and Online Dictionary Learnning (ODL)

	Sparsity level (L)	CDL	KSVD ($\gamma = 0.18$)	KSVDs (Atom sparsity level 10)	ODL
SNR 10 dB	2	**87.3**	72.8	61.20	86.47
	3	87.80	71.20	55.73	**88.69**
	4	**68.60**	28.53	38.93	65.16
SNR 20 dB	2	**90.67**	81.40	81.70	88.17
	3	91.67	82.50	75.67	**92.50**
	4	**93.60**	81.79	72.53	89.20
SNR 50 dB	2	90.67	**90.80**	88.57	89.52
	3	**91.67**	88.69	87.12	90.11
	4	**93.60**	91.55	90.54	87.42

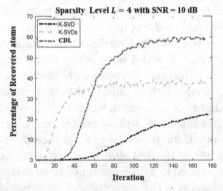

Fig. 3. Average percentage of atoms recovered after each iteration.

The proposed algorithm, Constrained Dictionary Learning (CDL), provides higher percentages of recovered temporal patterns than the three competitor algorithms (Table 1). This was especially true at lower SNR levels and at lower sparsity levels. The recovery rates of ODL were similar to the proposed algorithm at higher sparsity levels. Figure 3 depicts the average percentage of recovered atoms as a function of iteration number (sparsity $s = 4$, SNR $= 10$ dB). The figure suggests that K-SVDs was allowed to run to convergence but still arrived at a poorer solution than the proposed method. K-SVD had a much lower recovery rate possibly because it needs many more iterations to converge[2].

3.4   Real Task fMRI Data

The real fMRI data set was acquired as part of a clinical trial among healthy young men [17] ages 18 to 39. Functional magnetic resonance imaging (fMRI) scans were acquired on a 3T MRI scanner (General Electric, 750W Discovery, 32-channel quadrature head coil) using a blood oxygen level dependent echo-planar imaging (BOLD-EPI) pulse sequence. The fMRI scans collected during execution of the Attention Network Task [18] were analyzed from 50 participants, with congruent and incongruent trial types providing the two experimental conditions represented as 0 and 1 in the temporal constraint δ after convolution with a canonical hemodynamic response function. Spatial patterns of regions known to have different activity between congruent and incongruent trials, based on published reports (*i.e.*, the anterior cingulate cortex or ACC; and lateral prefrontal cortex or LPC) [19], were obtained from Automated Anatomical Atlas 3 (AAL3) [20]. These spatial patterns were used as the spatial constraints M in Eq. (3). A classical fMRI preprocessing pipeline of head motion correction, co-registration to the structural scan, slice timing correction, spatial smoothing using a 6 mm full width at half maximum Gaussian kernel, and warping to the Montreal Neurological Institute (MNI) template was applied using Statistical Parametric Mapping 12 (SPM12) with MATLAB R2020b. Participants wore a pulse oxygenation sensor during scanning, and cardiac/respiratory influences on fMRI signals were removed based on its data [21]. Data from the 20 participants with highest task performance accuracy was selected for analysis.

Dictionary Learning: Similar to the process carried out for the synthetic data, the data matrix was formed by the vectorized preprocessed images. The data matrix $X \in \mathbb{R}^{T \times n}$, was then down sampled by a factor of 8 along the column direction. A dictionary of size $T \times 40$ was initialized using random vectors (columns) taken from X. The same algorithms tested on synthetic data were applied to real data with sparsity level s set to 3. During optimization we constrained one time course δ (corresponding to congruent and incongruent trials) and one spatial pattern M (corresponding to executive control regions) and examined our algorithm's ability to learn spatial patterns corresponding to other FBNs known to be involved in task execution (the alerting and orienting networks) as well as temporal patterns corresponding to other relevant task conditions (*i.e.*, task trials with differing alerting and orienting requirements). As for synthetic data, the Pearson correlation between estimated and ground-truth temporal patterns were calculated, and

---

[2] The convergence rate of ODL is not included in the figure since the code did not provide this information.

the maximum correlation for each ground-truth pattern is reported as an indicator of recovery. Similarly, the Dice coefficient between estimated and ground-truth spatial patterns is reported as an indicator of spatial pattern recovery.

Table 2. Results for real fMRI data

Contrast	(A) Highest correlation coefficient between estimated temporal pattern and time course of task conditions				(B) Highest dice coefficient between recovered spatial pattern and known task-evoked networks			
	CDL	KSVD	KSVDs	ODL	CDL	KSVD	KSVDs	ODL
Executive control	**0.96**	0.53	0.61	0.68	**0.81**	0.59	0.52	0.45
Alerting	**0.83**	0.58	0.70	0.52	**0.65**	0.61	0.42	0.58
Orienting	**0.89**	0.47	0.39	0.48	**0.69**	0.51	0.32	0.51

Recovery of spatial and temporal patterns from the real fMRI data is shown in Table 2. As expected, our CDL method recovers the executive control spatial and temporal patterns because these were provided as constraints to the optimizer. Additionally, CDL does the best job of all algorithms at recovering the alerting and orienting related spatial and temporal patterns even though information about these spatial and temporal patterns were not provided to the optimizer. Interestingly, although the algorithm was encouraged to find a spatial pattern matching the spatial pattern of an executive control network, the spatial pattern providing the best such match was not perfectly correlated with the ground-truth spatial pattern (Pearson's $r = .81$), suggesting that requiring spatial patterns to exactly match ground-truth [6] may be too rigid to allow accurate data representation.

4   Conclusion

Using task information to partially constrain dictionary learning allowed this method to identify task related networks in real and synthetic fMRI data. Future work should assess how additional constraints help or hurt performance and how the method can be applied to more novel tasks with involved networks that are less certain. Standard online optimizer enhancements should also be explored such as 1) sampling the training data intelligently, 2) using Minibatch extension, and 3) purging the dictionary of unused atoms.

Acknowledgment. Funding for this project was provided by Collaborative Research to Optimize Warfighter Nutrition projects I and II, and the Defense Health Program, Joint Program Committee-5, Military Operational Medicine Research Program. Additional support provided by the Pennington Biomedical Research Foundation. Authors have no conflicts of interest to report.

Disclaimer. The views expressed in this paper are those of the authors and do not reflect the official policy of the Department of Army, Department of Defense, or U.S. Government. Any citations of commercial organizations and trade names in this report do not constitute an official Department of the Army endorsement or approval of the products or services of these organizations.

References

1. Elliott, M.L., et al.: General functional connectivity: Shared features of resting-state and task fMRI drive reliable and heritable individual differences in functional brain networks. Neuroimage **189**, 516–532 (2019)
2. Lo, O.-Y., et al.: Gait speed and gait variability are associated with different functional brain networks. Front. Aging Neurosci. **9**, 390 (2017)
3. Zhang, Z., et al.: A survey of sparse representation: algorithms and applications. IEEE Access **3**, 490–530 (2015)
4. Zhao, L., et al.: A task performance-guided model of functional networks identification. In: 2019 IEEE 16th International Symposium on Biomedical Imaging (ISBI 2019). IEEE (2019)
5. Olshausen, B.A., Field, D.J.: Sparse coding of sensory inputs. Curr. Opin. Neurobiol. **14**(4), 481–487 (2004)
6. Zhao, S., et al.: Supervised dictionary learning for inferring concurrent brain networks. IEEE Trans. Med. Imaging **34**(10), 2036–2045 (2015)
7. Jeong, S., et al.: Sparse representation-based denoising for high-resolution brain activation and functional connectivity modeling: a task fMRI study. IEEE Access **8**, 36728–36740 (2020)
8. Moreno, M.M., et al.: Assisted dictionary learning for FMRI data analysis. In: 2017 IEEE International Conference on Acoustics, Speech and Signal Processing (ICASSP). IEEE (2017)
9. Mairal, J., et al.: Online dictionary learning for sparse coding. In: Proceedings of the 26th Annual International Conference on Machine Learning. ACM (2009)
10. Morante, M., Kopsinis, Y., Theodoridis, S.: Information assisted dictionary learning for fMRI data analysis. arXiv preprint arXiv:1802.01334 (2018)
11. Aharon, M., Elad, M., Bruckstein, A.: K-SVD: An algorithm for designing overcomplete dictionaries for sparse representation. IEEE Trans. Signal Process. **54**(11), 4311–4322 (2006)
12. Lee, H., et al.: Efficient sparse coding algorithms. In: Advances in Neural Information Processing Systems (2007)
13. Mairal, J., et al.: Online learning for matrix factorization and sparse coding. J. Mach. Learn. Res. **11**, 19–60 (2010)
14. Bottou, L., Bousquet, O.: The tradeoffs of large scale learning. In: Advances in Neural Information Processing Systems (2008)
15. Mairal, J., et al.: Spams: a sparse modeling software, v2.3 (2014). http://spams-devel.gforge.inria.fr/downloads.html
16. Erhardt, E.B., et al.: SimTB, a simulation toolbox for fMRI data under a model of spatiotemporal separability. Neuroimage **59**(4), 4160–4167 (2012)
17. Pasiakos, S.M., et al.: Physiological and psychological effects of testosterone during severe energy deficit and recovery: a study protocol for a randomized, placebo-controlled trial for Optimizing Performance for Soldiers (OPS). Contemp. Clin. Trials **58**, 47–57 (2017)
18. Jennings, J.M., et al.: Age-related changes and the attention network task: an examination of alerting, orienting, and executive function. Aging Neuropsychol. Cogn. **14**(4), 353–369 (2007)
19. Matsumoto, K., Tanaka, K.: Conflict and cognitive control. Science **303**(5660), 969–970 (2004)
20. Rolls, E.T., et al.: Automated anatomical labelling atlas 3. Neuroimage **206**, 116189 (2020)
21. Kasper, L., et al.: The PhysIO toolbox for modeling physiological noise in fMRI data. J. Neurosci. Methods **276**, 56–72 (2017)

Author Index

Printed in the United States
by Baker & Taylor Publisher Services